Vulnerability of Coastal Ecosystems and Adaptation

From the ***Seas and Oceans*** Set
coordinated by
André Mariotti and Jean-Charles Pomerol

Vulnerability of Coastal Ecosystems and Adaptation

Edited by

André Monaco
Patrick Prouzet

ISTE

WILEY

First published 2014 in Great Britain and the United States by ISTE Ltd and John Wiley & Sons, Inc.

ISTE Ltd
27-37 St George's Road
London SW19 4EU
UK

www.iste.co.uk

John Wiley & Sons, Inc.
111 River Street
Hoboken, NJ 07030
USA

www.wiley.com

Library of Congress Control Number: 2014953025

British Library Cataloguing-in-Publication Data
A CIP record for this book is available from the British Library
ISBN 978-1-84821-704-1

Contents

Foreword

We have been asked by ISTE to stimulate work in the area of the environment. Therefore, we are proud to present the "Seas and Oceans" set of books, edited by André Monaco and Patrick Prouzet.

Both the content and the organization of this collection have largely been inspired by the reflection, initiatives and prospective works of a wide variety of national, European and international organizations in the field of the environment.

The "oceanographic" community, in France and internationally – which is recognized for the academic quality of the work it produces, and is determined that its research should be founded on a solid effort in the area of training and knowledge dissemination – was quick to respond to our call, and now offers this set of books, compiled under the skilled supervision of the two editing authors.

Within this community, there is a consensus about the need to promote an interdisciplinary "science of systems" – specifically in reference to the Earth's own "system" – in an all-encompassing approach, with the aim of providing answers about the planet's state, the way it works and the threats it faces, before going on to construct scenarios and lay down the elementary foundations needed for long-term, sustainable environment management, and for societies to adapt as required. This approach facilitates the shift of attention from this fundamental science of systems (based on the analysis of the processes at play, and the way in which they interact at all levels and between all the constituent parts making up the global system) to a "public" type of science, which is finalizable and participative, open to decision-makers, managers and all those who are interested in the future of our planet.

In this community, terms such as "vulnerability", "adaptation" and "sustainability" are commonly employed. We speak of various concepts, approaches or technologies, such as the value of ecosystems, heritage, "green" technologies, "blue" chemistry and renewable energies. Another foray into the field of civilian science lies in the adaptation of research to scales which are compatible with the societal, economic and legal issues, from global to regional to local.

All these aspects contribute to an in-depth understanding of the concept of an ecosystemic approach, the aim of which is the sustainable usage of natural resources, without affecting the quality, the structure or the function of the ecosystems involved. This concept is akin to the "socio-ecosystem approach" as defined by the Millennium Assessment (http://millenniumassessment.org).

In this context, where the complexity of natural systems is compounded with the complexity of societies, it has been difficult (if only because of how specialized the experts are in fairly reduced fields) to take into account the whole of the terrestrial system. Hence, in this editorial domain, the works in the "Seas and Oceans" set are limited to fluid envelopes and their interfaces. In that context, "sea" must be understood in the generic sense, as a general definition of bodies of salt water, as an environment. This includes epicontinental seas, semi-enclosed seas, enclosed seas, or coastal lakes, all of which are home to significant biodiversity and are highly susceptible to environmental impacts. "Ocean", on the other hand, denotes the environmental system, which has a crucial impact on the physical and biological operation of the terrestrial system – particularly in terms of climate regulation, but also in terms of the enormous reservoir of resources they constitute, covering 71% of the planet's surface, with a volume of 1,370 million km^3 of water.

This set of books covers all of these areas, examined from various aspects by specialists in the field: biological, physical or chemical function, biodiversity, vulnerability to climatic impacts, various uses, etc. The systemic approach and the emphasis placed on the available resources will guide readers to aspects of value-creation, governance and public policy. The long-term observation techniques used, new techniques and modeling are also taken into account; they are indispensable tools for the understanding of the dynamics and the integral functioning of the systems.

Finally, treatises will be included which are devoted to methodological or technical aspects.

The project thus conceived has been well received by numerous scientists renowned for their expertise. They belong to a wide variety of French national and international organizations, focusing on the environment.

These experts deserve our heartfelt thanks for committing to this effort in terms of putting their knowledge across and making it accessible, thus providing current students with the fundaments of knowledge which will help open the door to the broad range of careers that the area of the environment holds. These books are also addressed to a wider audience, including local or national governors, players in the decision-making authorities, or indeed "ordinary" citizens looking to be informed by the most authoritative sources.

Our warmest thanks go to André Monaco and Patrick Prouzet for their devotion and perseverance in service of the success of this enterprise.

Finally, we must thank the CNRS and Ifremer for the interest they have shown in this collection and for their financial aid, and we are very grateful to the numerous universities and other organizations which, through their researchers and engineers, have made the results of their reflections and activities available to this instructional corpus.

André MARIOTTI
Professor Emeritus at University Pierre and Marie Curie
Honorary Member of the Institut Universitaire de France
France

Jean-Charles POMEROL
Professor Emeritus at University Pierre and Marie Curie
France

1

Marine Ecosystems under Toxic Pressure

1.1. Introduction

In terms of the pressures of anthropogenic origin to which marine ecosystems are subjected, the ideas that spring to mind are intensive fishing, indirect forms of destruction, the destruction of habitats by fishing equipment during the exploitation of the deep sea, the development of ports – eutrophication, plastic macrowaste, etc. On the subject of toxic pressure, we can also mention the incidents of accidental pollution, of which the explosion of the oil-rig Deepwater Horizon, in the Gulf of Mexico in February 2013, recently gave us a sad example, or even, chemical shipwrecks, such as that of the Ievoli Sun in October 2000.

These events are indeed particularly striking because of the extent of the immediate mortalities that they cause, which are generally short-lived. However, the toxic pressures that marine ecosystems undergo due to the chronic and ubiquitous contamination of the environment by multiple contaminants are less well known and understood by the general public.

The first research on metallic trace elements in marine environments dates from the 1970s, and belongs mostly to the studies

Chapter written by Véronique LOIZEAU and Marie-Hélène TUSSEAU-VUILLEMIN.

of geochemists, searching to understand the global cycle of the elements of the Earth's crust, including some toxic metals, such as mercury, lead and cadmium and other essential metals, such as zinc and copper (see the reference work *Tracers in the Sea* by Broeclker and Peng [BRO 82]). Nevertheless, the goal, above all, is then to arrive at a precise knowledge of the quantity of these metals in seawater, which contains so little of them that handling them carries high chances of contamination... These metals are not, therefore, being studied as toxic pollutants. The emergence of ecotoxicology (the term was coined for the first time by René Truhaut in 1969) will very gradually lead to an understanding at least of the local effects in the sea, when accidents occur, or indeed in the vicinity of pollutant refuse. However, it was only with the application of the 1996 law on water that treatment plants emptying waste into the sea saw themselves obliged to control their emissions. Until then, it was thought that the oceans' power to dilute waste was sufficient protection. Finally, more recently, developments in analytical environmental and organic chemistry allowed the detection, in all environments, of organic xenobiotic[1] substances with toxic effects. Environmental chemists and ecologists became aware of the fact that the ocean, the environment where terrestrial life originated (see Chapter 1 of [MON 14c] also from the Seas and Ocean set of books), but which mankind does not inhabit, had nevertheless absorbed manmade chemical emissions and that marine ecosystems were living with this chronic pressure.

Since then, the major challenge facing scientists has been to understand how marine ecosystems behave under toxic pressure, what evolutions and adaptations this pressure causes, and at what cost (metabolic, phenotypic and genetic). In effect, other major pressures are being exerted, among them climate change and acidification of the oceans, etc. (see [MON 14a, b and c] also from the Seas and Oceans set). And the future of the entire biosphere is directly linked to the oceans' capacity to sustain significant primary production, trapping atmospheric carbon dioxide, which is the basis whole of the

1 A substance present in a living organism, which is nevertheless foreign to it.

trophic oceanic chain, which feeds not only the sea birds... but also people.

First, these are the details of the ocean environment that are affected *vis-à-vis* the toxic pressure. Then, the biological responses will be described at the level of individuals exposed to toxic pressure, independent of each another ("direct effects"). Finally, the focus turns to the group of effects known as "indirect effects", that is to say, those that affect the relationships between the individuals of which an ecosystem is composed. Little is yet known about these indirect effects, but initial observations have tended to show that they are the primary impact; understanding the behavior of these systems under toxic pressure has to be taken into account.

1.2. Details of the marine environment

All aquatic environments are subject to pollution of anthropogenic origin, and all the associated ecosystems are subject to the toxic stress that results. The ocean, because of its dimensions – it is the most vast of the biosphere's ecosystems (1.4 billion km^3) whose depth reaches, on average 3,800 m – and because of its distance from the continents appears relatively protected in comparison to rivers and lakes. Rivers and lakes are often very directly impacted by human use: runoff from agricultural land or soil that has been made impermeable, sources of diverse phytosanitary products, hydrocarbons, dioxins, metals, etc.; they are the recipients of more or less well-treated collection networks, sources of molecules from pharmaceutical synthesis, cosmetics, detergents, products from eroded materials, etc.; outlets, finally, of the widespread contamination of our environment by extremely varied products (see also Chapters 2 and 3). Locally, the impacts of these contaminations can be very pronounced (for example [DED 09], chemosphere), even if they are difficult to prove, because of the mobility of flowing water [FEC 14] and the physico-chemical variability of these environments: diurnal variations in pH and temperature, seasonal variations in organic matter and in shade from forest cover, regional variations in the concentration of eroded minerals, etc., all are modulating factors in the bioavailability of the contaminants [TUS 07].

Assessment of the contamination of the marine environment – which is vast, chemically and thermically stable, and relatively homogeneous in the oceanic areas – and of the consequences for the associated ecosystems is, therefore, fundamentally different. In fact, it is important to distinguish coastal environments from open oceanic environments, situated beyond the continental plateau. The risks of contamination in the coastal zone, to which estuaries and laguna can be added, are fairly similar to those of continental environments, down to a few specific details.

1.2.1. *The coastal zone*

This interface between the continent and oceans is home to specific ecosystems where important transfers of matter, energy and genes occur. The marshes, the seasonal nature of rivers' hydrology and the pre-eminence of primary production confer on the coastal zone a physico-chemical instability analogous to that of continental waters. The biodiversity housed by coastal regions is adapted to the strong variability in the characteristics of these transition environments, but its resilience has been broken down by anthropic impacts, leading to an increased vulnerability to pollution and global change, even more critical in the case of islands and lagoons. The recipient of nutritive salts eroded or washed from continents, the coastal and littoral zone, which is not very deep, provides numerous services to ecosystems (support and regulation especially) via primary production, the recycling of major elements, the metabolization of contaminants or their export into sediments and hydrological regulation. Costanza *et al.* [COS 97] estimate that a third of the global benefits and services to ecosystems are formed there. Because of this, but also due to the access to waterways that they provide, as well as the attraction that they exercise for our contemporaries, coastal areas concentrate 60% of the world's population – which is becoming increasingly urban and concentrated in megacities – at least 100 km from the coasts.

Coastal ecosystems, rich and vulnerable by nature, are therefore subject not only to pressures provoked by global changes, including climate change, but also to pressures due to this very strong concentration of continental activities as well as maritime activities.

The species exploited (fishing, conchiliculture (the farming of shellfish), the farming of sea vegetation, etc.) are also subject to these pressures, which could explain certain recurrent weaknesses in the immune system (F. Akcha, personal communication). Moreover, exposure to pollutants can lead to a contamination of the biomass, rendering it unfit for human consumption [LEB 06].

Once they have passed through the filters of lagoons, deltas and estuaries, in which the levels of salinity trigger a significant precipitation of matter, trapping certain components (cadmium in the Gironde or in the Bay of the Seine, for example, among the most well-known instances [SHI 13]), a proportion of the collection of micropollutants issuing from the drainage basin are found in the coastal zone (hydrocarbons, pesticides, metals, persistent organic products, medication, cosmetics, etc.). For example, the supplies of hydrocarbons to the marine environment account for 80% of the telluric supplies, accidental pollution, therefore, only represents a small fraction, of which the impact is mainly local and often significant for macrofauna. At the end of the 1980s, the development of chromatography in liquid form for environmental research, coupled with mass spectrometry at high resolution, enabled hydrophilic molecules of pharmaceutical, cosmetic or hygienic origin to be gradually detected, at weak concentrations that qualify as emergent. Pioneering studies have enabled the identification of a number of these substances active in rivers, lakes and aquifers (for example, in the United States [KOL 02]). In France, the first studies on coastal waters and estuaries were only carried out quite recently [CAS 06]. Antibiotic and anti-inflammatory products, fungicides, antidepressants, analgesics and anticancer medications have been identified, of which it is still difficult to evaluate the real impact on fauna and aquatic flora. More specifically, the use of the coast leads locally to strong concentrations of cosmetic sun protection products, oils and perfumes, as well as products used to protect the hulls of boats from biological fouling (copper, tributyltin (TBT) and its replacements). For example, at only 20 $ng.L^{-1}$, TBT considerably disrupts the growing metabolism of mussels. At 2 $ng.L^{-1}$, TBT, an endocrine disruptor, is capable of modifying the sex of certain marine gastropods (masculinization of the females by the effect called imposex) [ABI 12]. Nanoparticles of titanium oxide, used in a lot of sun protection products, are part of the emerging concerns. Nevertheless, in their review on the subject, [KLA 09] do not report any more observations *in situ*. Research is currently concentrated on the evaluation *in vitro* of the potential effects of these nanoparticles on coastal organisms [CAN 10]. Finally, the coastal zone permits active exchanges between the column of water and the sediments, which unfortunately often constitute a significant reservoir of persistent contaminants. Much like continental lakes, the toxic threat associated with them thus lasts decades, being reactivated following the reshaping of the sediment or modifications of its "redox status".

1.2.2. *The open ocean*

The risks from contamination of the open ocean by toxic substances are different. There is less biodiversity there, which results from the low habitat fragmentation of the environments, and from their relative homogeneity. Nevertheless, marine ecosystems are of the greatest importance for mankind: for their supply of protein biomass (15% of the total supply) and for their value to local communities: the cradle of life on Earth, still largely unexplored, is believed by some to be outside the reach of anthropogenic pressures [GOU 12, p. 18]. They are home to wild species of great longevity, at the top of the trophic chain.

One way of taking account of the immensity of the oceans and their inertia consists of evaluating the residence time of elements in their different compartments. This amount, homogeneous for a time, is obtained by dividing the volume of a reservoir by the fluxes that cross it, under the hypothesis of stationarity. The residence time of water in the global ocean is in the order of 3,000 years [DE 09]. For the Mediterranean, it is, for example, a hundred years. This means that an easily biodegradable contaminant, carried from the continent, such as glyphosate (in the order of a month in water, INERIS, 2010), will not be found again in a significant concentration in the whole of the

Mediterranean basin. However, persistent organic contaminants, such as PolyChlorinated Biphenyls (PCB), can easily be dispersed. In effect, the duration of the half-life in water (suggested by [MAC 92] to be two years for tri- and tetra-chlorides, and six years for penta- to hepta-chlorides) does not include transfers in the food chain, one of the most efficient methods of storage and transport for hydrophobic substances. Chlordecone, a chlorine insecticide used to combat the banana-tree weevil in the Antilles, is one of the contaminants for which it is still difficult to suggest a typical biodegradation time. In fact, the risks for the open ocean, where ecotoxicology is concerned, are on the one hand those of persistent contaminants on the scale of oceanic fluxes (a decade and more), and on the other hand of substances whose planetary cycle is in part controlled by specific marine processes. Mercury can be counted among the latter, of which the atmospheric supplies through snow, then the arrival at the ice interface of sea and seawater appears to be primordial [COS 11, DAS 14], but whose planetary cycle remains to be elucidated.

These compounds are generally hydrophobic – this is what makes them difficult to biodegrade in an aquatic environment – and lipophilic, and they, therefore, spread in the trophic chain in spite of the fairly weak concentrations that cause them to be diluted in the ocean. Their enduring presence, associated with this particular mode of transfer, leads large sea predators to become contaminated, whether these are, for example, tuna [KRA 03], mammals or even birds [DIE 13]. The risks are, therefore, both the chronic toxicity of these substances for large organisms and the ecosystems to which they belong – which for a long time were thought to be unable to carry pollution of human origin – as well as the fitness of the human foodstuffs that are taken from them. This concern is even more important when a population's food supply is mainly taken from marine sources, as is the case, for example, for the Inuit [DAL 13] and Polynesians [DEW 08].

More than 37 million chemical substances are currently listed in the world, for the most part substances resulting from biosynthesis. Around 100,000 chemical substances are produced, imported and used on the European market, and 5,000 of them (5%) are considered to be dangerous for mankind and the environment. Sources of contamination by metals are multiple and include mining activity, the steel industry, transport, the use of different types of batteries and the painting and dying industries, as well as the use of phosphorous fertilizer (cadmium). Taking account of the diversity of the molecules, the study of organic contaminants represents a very important undertaking. Very schematically, it is possible to distinguish four main substance groups:

– hydrocarbons, of which aromatic polycyclic hydrocarbons (APHs) are the most worrying for aquatic environments;

– pesticides, with some 900 types in current use and a usage rate of 80,000 tons applied each year;

– biocides, which refers to substances used in a non-phytopharmaceutic context, such as additives included in anti-fouling paints for use at sea, which cause non-negligible contamination by different organometallic (TBT) or organic (diuron or atrazine such as Irgarol 1057) or active metallic substances (copper);

– other organic synthetic substances that represent a large number of substances (chlorinated solvents, PCB, flame retardants, phthalates, detergents, colorants, etc.). The selection criteria for chemical contaminants judged to be a priority for the environment are based on three properties: persistence (P) defining persistent substances in the environment (for example, persistent organic pollutants (POPs) such as DichloroDiphenylTrichloroethane (DDT)), bioaccumulation (B) defining their capacity to accumulate in organisms and toxicity (T). These three properties define a group of substances that are called PBT substances. To this group should be added substances that have carcinogenic and mutagenic properties and effects on the reproductive system that are called CMR substances. Endocrine disruptors are also associated with this group.

Box 1.1. *The chemical universe and the environment*

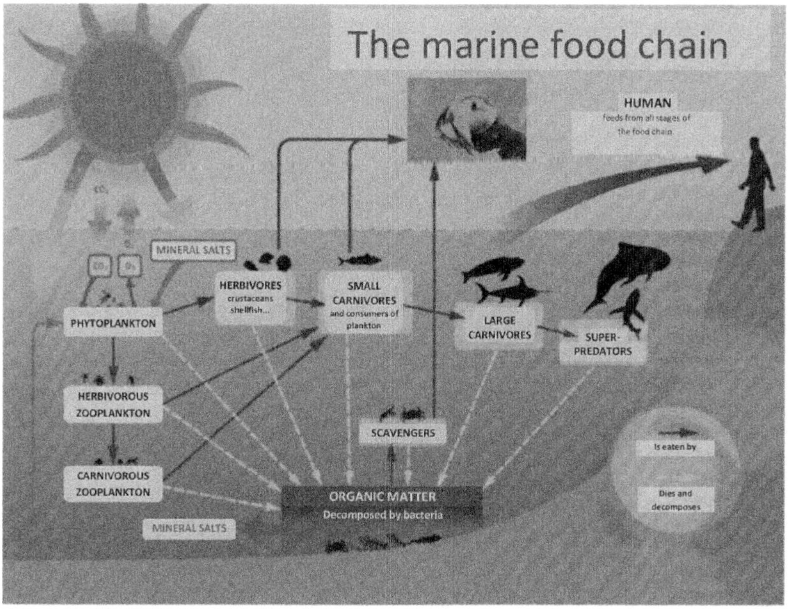

Figure 1.1. *Example of the marine trophic chain*

1.3. What is the biological response of organisms to contaminants?

The absorption of a contaminant by a living organism triggers a disruption in its metabolism. This disruption leads to a biological response, which results either in the cell returning to its non-disturbed state or in a manifestation of toxic effects. An organism's biological response to a toxin can be seen as the result of the interaction between the intrinsic properties of the substance and those of the organism exposed to it. It depends on the chemical properties of the contaminant (structure, activity and mode of action) and the physiological, biological and ecological properties that characterize the organism at the moment in its life when it is exposed to this contaminant. It is also variable within a single species and a single age group, depending on individuals. For example, many of us are exposed to the flu virus each winter, and only some will actually become ill. Its general physiological state has, in effect, an impact on the reaction of an organism in the face of a stressor. Each species, and each individual within a species, therefore shows a specific response to each chemical product. Moreover, the environment in which this species evolves and its connection with other parts of the ecosystem will also condition its response. Finally, the direct impact of contaminants on a species and/or a subsection of it can generate indirect effects on the entirety of the ecosystem, as we will see later (Figure 1.2).

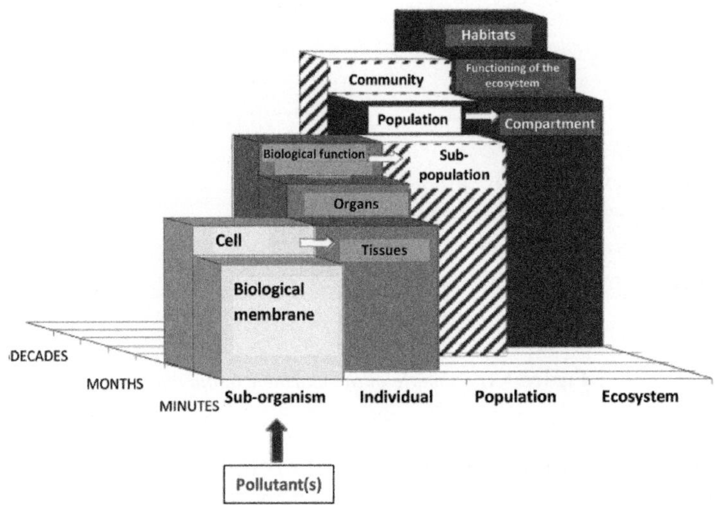

Figure 1.2. *Adaptation of the schematic representation of the field of ecotoxicology studies depending on the level of complexity of the lifeform and the time, from [MUN 95] and [ADA 00]*

1.3.1. *At cellular level*

The manifestation of a disturbance in the functioning of the cell is the result of interaction between the cellular biomolecules and contaminants. This interaction is very specific: its occurrence and intensity depend on the organism's physiology at the moment it is exposed and on the toxic product's mode of action.

1.3.1.1. *General remarks on the modes of action*

The contaminant's mode of action is determined by its chemical structure [ESC 02, PAK 00, TRE 04]. Pollutants have been classified into four categories, on the basis of their structure/activity ratios (SARs) [HER 89]. The "inert" products (class I) and "slightly reactive" products (class II) have a narcotic action: they react in a non-selective manner with biological membranes, thus modifying their structure and functioning. Their effect depends mainly on their hydrophobia [VAN 92]. The "reactive" compounds (class III) and the compounds "with a specific mode of action" (class IV) react selectively or not at all with the cellular biomolecules. Once in the cell, they are generally hydroxylated and eventually combined with other molecules in order to be eliminated. However, the combinations are sometimes more toxic than the initial substance.

For a family of contaminants with a given mode of action, the cell's response depends on the presence and the abundance of the product's targets and on its metabolic capacities, as well as on its ability to repair damage [ESC 02]. These physiological properties characterize species. For example, the active elimination of the contaminant by cells calls upon different metabolic paths whose biomolecules are unequally distributed between different species [BAZ 97, CAL 83, IBR 98] or within the same species. Thus, a given cell's membership of a taxonomic group determines its response to a given product.

1.3.1.2. *The cellular response: the means of identifying exposure to contaminants before the event*

The impact of pollutants on a subcellular level can lead to the inhibition and/or triggering of diverse proteins and enzymes implicated in the metabolism and the excretion of xenobiotics. These detoxification mechanisms allow organisms to maintain themselves in the fact of exposure to pollutants. Modulations of biotransformation enzymes have, therefore, been the subject of a very large number of investigations over the last 30 years, notably among fish [AND 92, GOK 98, WHY 00]. Much effort has, in particular, been devoted to the identification of biomarkers of detoxification, that is to say proteins, or indeed enzymes, whose activity levels reveal the starting of processes within the cell. This work focuses on the measurement of cytochrome protein levels P450 (phases 1A and 3A) [MUR 97, WEB 02], the measurement of ethoxyresorufin-O-deethylase activity (EROD) [GOK 98, TEL 04, WHY 00] or the enzymes from the glutathione-S-transferase family [GEO 94, KIM 10, VAR 89]. As an example, Figure 1.3 shows a correlation between levels of contamination and EROD activity [BUR 94, GAL 91]. These molecular biomarkers are not, however, specific to the contaminants that trigger their activation, and their responses are potentially affected by biotic or abiotic factors.

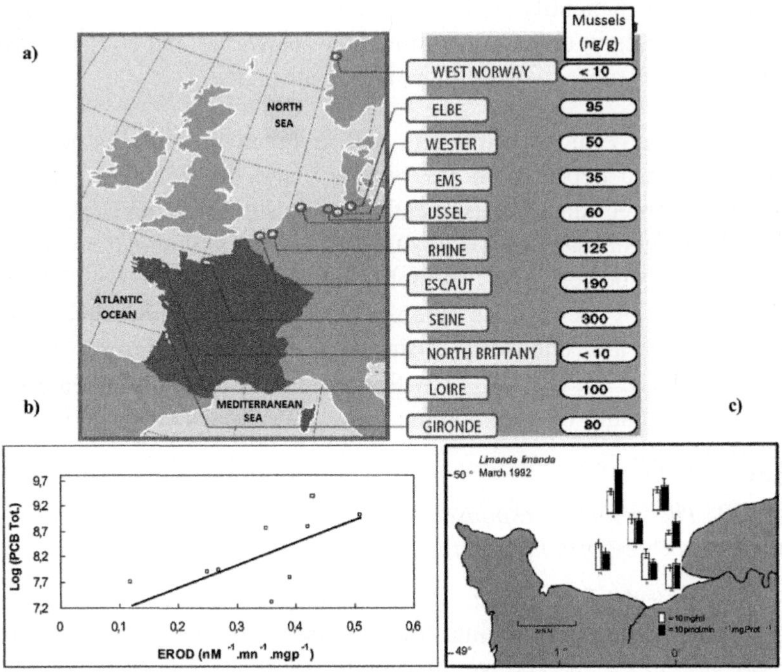

Figure 1.3. *a) Contamination in mussels (ng.g^{-1}) on European coasts by PCB from measurements in RNO 1991; b) correlation between EROD activity (nMol produced by enzymatic activity (resorufin), per minute and per mg of liver protein) and level of contamination in PCB (ng.g^{-1}) in samples of plaice [GAL 91] and c) EROD activity (by the same units as (b)) in samples of flounder in the Bay of the Seine [BUR 94]*

1.3.2. *On an individual level*

Disturbance in cellular physiology generally manifests itself by the effects on the survival, growth, reproduction and indeed the behavior of individuals. It is also at an individual level that contamination occurs. This does not depend only on the chemical form of the contaminants, from which it has wrongly been thought possible to define "the" bioavailability [GOU 13], but above all on the specific details of an individual's life history.

1.3.2.1. *How specific and individual variability influences contamination*

The biological and ecological characteristics of individuals are implicated at all stages of the contamination process and the biological response: exposure, absorption, elimination and eventual compensation for the product's effects. In the first place, the duration of contact between the organism and contaminated environment depends on the number and duration of the developmental stages undergone during the lifecycle, as well as the presence of defense mechanisms over the course of this cycle [SPR 05]. The habitat and feeding method determine the organism's behavior and influence its level of exposure [KOI 92].

The biological and ecological characteristics of species are moreover involved in the kinetics of the organism's contamination [ESC 02]. In effect, the speed at which a contaminant is absorbed depends on the intensity of the exchanges between the organism and its environment. This absorption speed can be described using food assimilation rates [CAN 02] and the exchange surface between the organism and environment. This exchange surface is generally represented by the ratio between the surface of the body and its volume (S/V) [ESC 02]. In this ratio, assessment of the body surface takes account of the toxin's different absorption routes: the integument, the digestive tract and the respiratory surfaces [WEI 04]. For two organisms of similar size, the higher the S/V ratio, the more rapid the kinetics of the toxin's absorption [KOI 92]. In practice, it is, therefore, mainly the mode of feeding and the respiratory system (gills or integument) that is involved in the organism's contamination kinetics.

1.3.2.2. *How specific variability and individual influence the depuration rate*

The biological and ecological characteristics of species and individuals are also involved in their capacity to eliminate or store the contaminant in a non-dangerous form. Passive elimination of the compound implies its excretion or accumulation in inert compartments of the organism [GRO 99]. Furthermore, tissues with a high lipid content offer a significant storage volume of hydrophobic contaminants. Differences in the presence and volume of these compartments generate a strong interspecies variability in biological response to toxins and, in particular, to organic products [ESC 02]. These differences are linked to the relative size of species: in effect, it is generally the largest organisms that possess the most lipidic reserves [CAN 02, ESC 02].

Thus, it is mainly the characteristics linked to use of the habitat and food, as much as the characteristics linked to the pattern of the organisms' life history, which determines their response to contaminants at an individual level.

1.3.2.3. *Some types of toxic effects*

Certain pollutants, characterized as "endocrine disruptors" (PE), act on organisms' hormonal equilibrium. Endocrine disruptors are exogenous substances that trigger effects harmful to the health of an organism or its descendents, following changes to endocrine function. The action mechanisms of PEs are multiple, since they can act on all the stages of endocrine regulation, from the synthesis of hormones to activity at the level of the target tissues. Among aquatic organisms, exposure to PEs has been associated with harmful effects on the reproduction (Figure 1.4; [CRA 08, DAO 11, MEN 08, MIL 05]) of individuals and populations. Certain pollutants have also been identified as neurotoxic, leading to effects on the neural functions of fish, and they can, therefore, potentially affect the species's behavior and learning [PEA 13].

Diverse chemical agents have the capacity to interact with the DNA molecule and to modify its nature [CAJ 03, HEB 96]. Types of damage to DNA are generally separated into two categories: genetic lesions (functional lesions) caused by mutagenous agents and chromosomic lesions (structural lesions) caused by clastogenic agents [EVE 94, LIV 00, MAR 10].

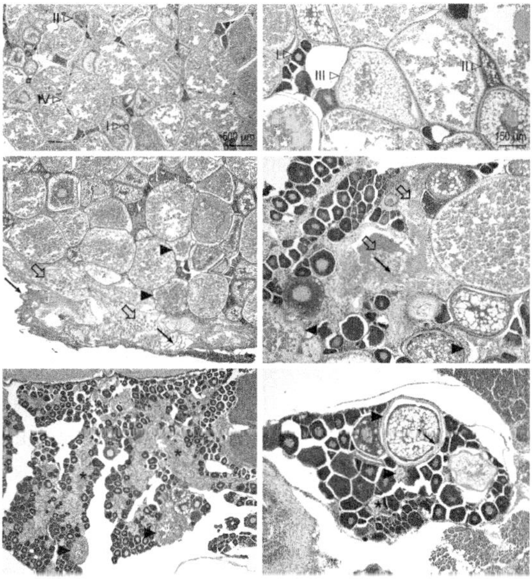

Figure 1.4. *Histological structure of the ovaries of female zebra fish (taken from [DAO 11])*

COMMENTARY ON FIGURE 1.4.– The different stages of maturity of ovarian follicles (I–IV) can be identified in the ovary of a control fish. The fish ovary exposed to PCBs (at levels in the order of those encountered in the Loire estuary; photo, center) shows a number of atretic (arrow) follicles. The number of follicles at stage III (vitellogenous) and stage IV (mature ovocite) is slightly diminished in comparison to those of the control. The ovaries of fish exposed to elevated concentrations of PCB (Seine estuary; photo below) show an almost total absence of follicular stages III and IV, which makes reproduction impossible.

1.3.3. *On the level of the population*

The decrease in rates of survival, growth and reproduction, as well as the modification of individual behavior – as far as it is due to them – leads to a modification in the population's dynamic, which can go as far as the disappearance of the species in a polluted habitat [CAS 01]. For a single disruption at an individual level, the effect on the population is lesser or greater according to the pattern of the species' life history [SPR 05].

In practice, the pattern of the species' life history is a group of data that includes: the species' biodemographic parameters in a given environment (lifespan, number of developmental stages, number of offspring, fecundity, etc.) and the ratios between these parameters (ratio between lifespan and age at which an organism first reproduces, ratio between rates of growth and fecundity, etc.). Each pattern of the species' life history corresponds to a coadaptation of the species' biological and ecological characteristics depending on the biotic and abiotic factors of its habitat. These characteristics can be classified into two categories. Those that are linked to the developmental cycle of the organism influence the probability of survival above all. Those that are linked to reproduction influence fecundity above all.

Thus, the number of developmental stages and the population's vulnerability in terms of the probability of survival at these stages are major determinants of a population's resistance to disruptions [SPR 05]. In effect, among certain organisms, there are critical periods or stages of development during which the organism's probability of survival is weaker than at other stages [KOE 91]. These stages are particularly vulnerable to instances of pollution. For example, this is the case during periods of larval molts among crustaceans [KOI 92] or the metamorphosis between the larval and juvenile stages of fish (for example, flat fish, which pass from a pelagic larval life stage to a benthic stage during their juvenile and adult phases).

The characteristics influencing organisms' reproduction, along with their lifespan, strongly influence a population's response to contaminants [CAL 97, IND 99]. In effect, the number of descendants produced per year in a population depends, according to [SPR 05]:

– on the age of sexual maturity compared to the lifespan and the number of potential reproductive opportunities;

– on the number of reproductions per year;

– on the organisms' fecundity (the more energy the parent invests in the egg, the less productive it is, but the higher the chance of survival of the egg, and then the juvenile);

– on the existence of parental care of the eggs and/or the young.

In a non-polluted environment, there is no "good" or "bad" strategy in terms of a population's persistence [SPR 05]. However, certain strategies are more "efficient" than others in terms of persistence when pollution is added to the normal level of disruption in the environment [KAM 96]. For example, in the case of a short and temporary disruption lasting for the duration of the reproductive season, species that can reproduce several times in the course of the season will probably be less affected than those that only reproduce once [CAL 97].

In conclusion, a cell's response to a contaminant results from the interaction between its physiology, function and activity, and the mode of action of the contaminant to which it is exposed. The biological response of an individual or a population results from the interaction between the disruptions caused by the toxin at a cellular level and the organism's biological and ecological characteristics. These characteristics are mainly determined by the properties of the habitat and the organism's life history strategy.

Nevertheless, specific populations are only one element of their communities and ecosystems. It is becoming much clearer that it is at this level of lifeforms' organization, when exchanges of matter, energy and genes occur, that modes of evolution, adaptation and resilience are made permanently, notably in response to toxic pressures. Figure 1.5, taken from [ADA 05a], shows the different levels at which environmental stress produces direct and indirect effects on the ecosystem's structure.

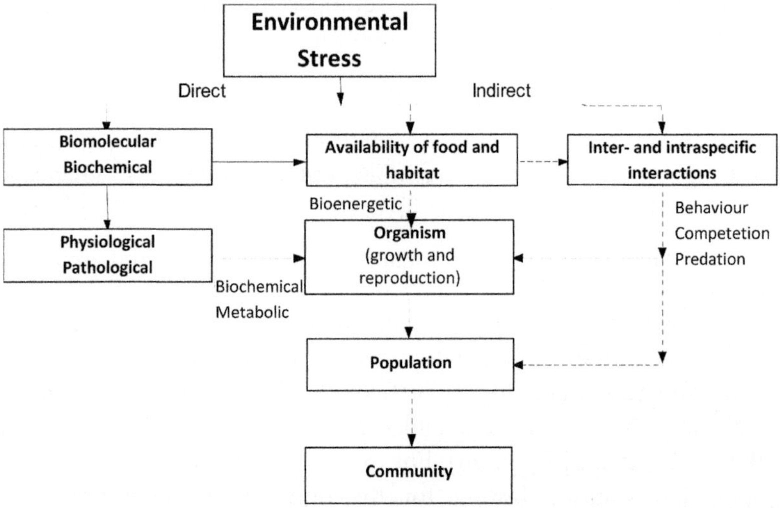

Figure 1.5. *Direct and indirect effects of environmental stress on biological systems (taken from [ADA 05a])*

The direct effects affect organisms by acting on the biochemical and metabolic processes; indirect effects disrupt the biota[2] via the availability of food and habitat, and via intra- and inter-species interactions.

1.4. Consequences of toxic pressure on ecosystems

When pollutants are ejected into aquatic ecosystems, direct (toxic) effects on biota are possible. These direct effects of toxic substances tend to reduce the number of organisms (by an increase in mortality and/or a decrease in fertility). The biota of a given habitat often show a large range of tolerance to specific toxic substances (for example, insecticides and herbicides target specific organisms in a community), the toxin can have lethal effects on certain species, but have no observable effect on others. Pollutants can, however, have effects on tolerant species through a number of ecological mechanisms. These

2 A group of living organisms present in a particular habitat or more generally in a given region.

effects are called indirect (or secondary) effects of the contaminants [PAL 08]. Toxicity tests in laboratories based on a single species do not permit indirect effects to be detected. To assess the impact of pollutants on the scale of a population, a community or an ecosystem, studies carried out in a microcosm or *in situ* are necessary [CAI 83, CLE 94].

The effect of contaminants on one section of an ecosystem (predators and/or prey) can lead to a cascade of indirect effects on species *a priori* resistant at other trophic levels. These effects on sensitive species can, therefore, modify concurrential interactions in resistant members, both producers and consumers, of the community. Furthermore, toxic substances can act directly on "key species" of an ecosystem. These species, including, for example, "engineer species" or "keystone species"[3] have an influence on other biological compartments via modifications in the environment [BRU 01]. Thus, mechanisms linked to the population and/or community after exposure to contaminants are potentially complex and very varied. The indirect effects of toxic products can then lead to an increase (reduced concurrence) or a decrease in the abundance (reduced availability of the "preferred food").

1.4.1. *Interspecies relationships*

At relatively high concentrations, the contaminants are fatal and cause indirect effects on the "density control" relationships between species [FLE 03, DE 89]. The contaminants, present at non-lethal concentrations, can also have an incidence on a broad range of individual biological traits (changes in the neurotransmitters, hormones, immune response and reproduction) and behavior (hunting for food, capacity to swim, detection of predators, learning and social interactions) [WEI 01]. These modifications of course affect the individual, but can also have indirect effects on the community in which the individuals are integrated.

3 Of which the disappearance leads to a modification, indeed to a brutal degradation of the ecosystem.

1.4.1.1. *Predator–prey interactions (top-down)*

The "top-down" effects are triggered when a predator is more sensitive to a contaminant than its prey (Figure 1.3). This sensitivity, without consequently directly causing the predator's death, can modify the ingestion and/or predation rates [GRE 97, WAL 00, WEI 01, WID 91] and thus causes an increase in the abundance of the prey (Figure 1.6). "Top-down" effects have thus been used as evidence in marine benthic systems after contamination by metals, fungicides and hydrocarbons [BEL 00, CAR 97, LAY 85, JAK 96, VAN 00]. These different contaminations have all led to trophic cascades. In addition to reducing pressure from predation, the contaminants can also benefit producers via the release of nutrients from the decomposition of deceased animals [KNA 05].

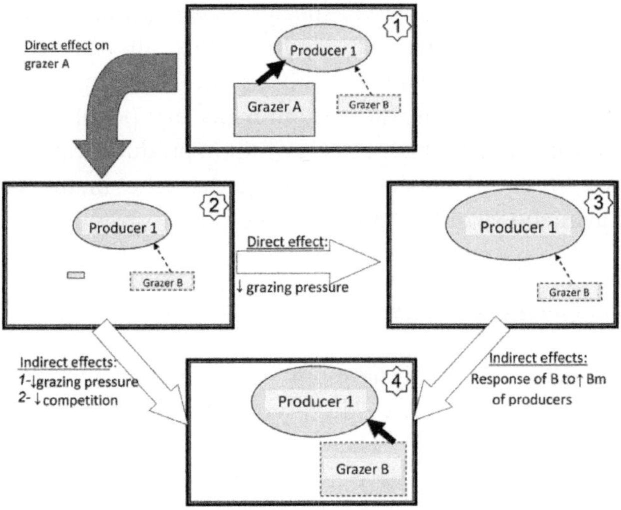

Figure 1.6. *Illustration of indirect "top-down" effects of contaminants (modified from [FLE 03])*

COMMENTARY ON FIGURE 1.6.– Box 1 represents a community of primary producers and two species of grazers (A = dominant; B = rare) before modification by a contaminant. Box 2 shows the result of a possible direct effect (rapid and selective) of the contaminant on the dominant grazers (A). The primary producers benefit from the reduction in pressure from grazing due to the death of (A). Boxes 3 and 4 illustrate two indirect effects able to lead to an increase in the abundance of grazers (B) (box 4). First of all, the "rare" grazers can benefit from the death of the common grazers and thus increase their rate of feeding on the primary producers through a decrease in the concurrence, which leads to an increase in the population. Box 4 shows that the increase in the abundance of primary producers generates a functional response for rare grazers, who become dominant.

The structure of the community has evolved from 1 to 4 following exposure to a contaminant.

1.4.1.2. *Interaction between resources and consumers: "bottom-up"*

Contaminants can have indirect effects at the base of the food chain (vegetables, detritic and/or bacterial biomasses, primary producers) by modifying, for example, the organic substrate of detritivores or by affecting the liberation rates of nutritive elements for primary producers. Hydrocarbons from petrol represent an important source of organic matter that can lead to a stimulation in productivity or the bacterial biomass. This stimulation can be sufficiently important to feed a reaction of the "bottom-up" type on bacteriophagic species and/or detritivores [PET 96]. In another domain, Podemski and Culp [POD 01] found evidence for a stimulation in the growth of diatoms subjected to effluent coming from a paper-whitening factory. This phenomenon generated an important proliferation of grazers feeding on the diatoms.

Herbicides can have direct effects on microalgae and generate indirect effects on communities of zooplankton via variations in competiveness between species. Through cascade effects, these are the species at higher trophic levels, such as salmon, which are more susceptible to being affected [BRO 10].

Fleeger *et al.* [FLE 03] list 56 cases of indirect effects from pesticides on the competition between species or predation on the biota and conclude that, at least in aquatic systems, pesticides exert a strong selection pressure on invertebrates.

Few studies on aquatic systems have examined the effects of contaminants on the cycle of nutritive elements or their bioavailability. Studies carried out on land ecosystems suggest, however, that the contaminants can increase or on the contrary decrease the liberation of nutritive elements at such a level that the abundance of the vegetation and/or the bacterial population can be affected [BOG 96, SAL 97, SAL 01].

1.4.1.3. *The alteration of behavior*

The behavior is defined as the group of coordinated internal responses (action or inaction) of living organisms (individuals or groups) to internal and/or external stimulae, excluding responses linked to development [LEV 09]. A behavior is, therefore, an interaction of living beings with each other or with their environment (Figure 1.7). It can appear to change or be inhibited following a change in physiological state, a change in the environment or due to a new social interaction. It is a very early and sensitive indicator of physiological disturbances. An individual response, alteration in behavior very often leads to indirect effects linked to predation.

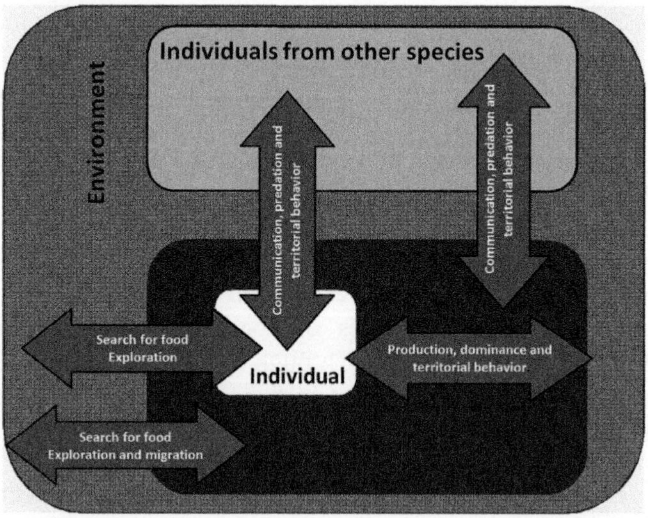

Figure 1.7. *Different behavioral relationships between an individual with the other individuals of its species, other species or its environment (taken from [PEA 12])*

The behavior of living organisms is susceptible to being affected by the presence of toxic substances in the environment [WEI 01]. Disturbances in organisms' locomotive activities can lead to deficiencies in their capacity to flee when faced with predators and an increase in the time required to search for prey as well as modifications in migratory activities indirectly affecting their reproduction [TRI 97, WEI 02, WEI 04]. Several studies have also indicated that pollutants can decrease the alimentary activity of fish. Finally, the presence of undesirable molecules can lead to modifications in behavior linked to reproduction (display, choice of partner and parental care). All these behavioral changes result from modifications in the biochemical and physiological processes on an individual level and can have significant effects on the equilibrium of populations and communities. In effect, the contaminants can cause behavioral or defensive responses, which can modify biological interactions or even intensify the effect of a contaminant.

Three general types of pollutant influence on the behavior have been observed:

– reduced rates of predation: although numerous studies demonstrate that ingestion or rates of predation by various animals can be diminished by contaminants, very few identify a behavioral mechanism associated with the hunt for food. Smith and Weis [SMI 97], however, have stated that exposure to mercury has been shown to correlate with a decrease in the predation intensity of the killifish (*Fundulus heteroclitus*); Temara *et al.* [TEM 99] have described changes in the predatory behavior of a starfish exposed to soluble fractions of crude oil;

– an increased sensitivity to predation: exposure to contaminants can inhibit a prey's specific behavior used to avoid a predator. This effect can lead to an increase in rates of encounters between predators and prey. As an example, Hinkle-Conn *et al.* [HIN 98] have found evidence of a decrease in several species of invertebrates' capacity to dig holes in contaminated sediments, increasing their exposure to predators. Many other examples of increased sensitivity to predation, triggered by a variety of contaminants, are identified in several studies in fresh and marine waters, involving both vertebrates and invertebrates at once [CLE 99, CLE 09, DOD 95, HAM 95, KIF 96, KRU 94, LEF 99, PRE 99, TAY 95, WEI 95, ZHO 99]. The increase in the intensity of predation can lead to an increase in the rates of trophic transfer from contaminants to contaminated prey to larger predators;

– a reduced sensitivity to predation: the changes caused by the contaminants on the behavior of prey that would decrease the probability of encounters between prey and predators, thus reducing the rates of predation, have received relatively little attention. However, Taylor *et al.* [TAY 95] have found evidence of a decrease in a cladoceran's capacity to swim in the presence of lindane. This disruption has reduced its vulnerability to predation by the hydra (*Hydra oligactis*).

1.4.1.4. *Interspecies competition within a same trophic behavior*

One of the clearest cases in which contaminants can affect interspecies competition can be found in the examples of differential sensitivity in several species within a single trophic group. For example, several studies have demonstrated that different groups of zooplankton, which are in competition for phytoplankton, have different sensitivities to insecticide (typically, cladocerans are more sensitive than copepodes). Thus, when moderate concentrations of insecticides are applied to communities containing cladocerans and copepodes, a spectacular drop in the more sensitive cladocerans is observed while the number of copepodes increases [HAV 93, HAV 95, VAN 02, REL 05]. Several authors have also examined the effects of insecticides on communities of macroinvertebrates composed of several competing species. Thus, populations of gastropods and oligochetes increase following exposure to a mixture of lindane and chlorpyrifos to the detriment of crustaceans more sensitive to these insecticides [CUP 02]. Differential sensitivity to herbicides has also been documented among producers. In a green-macrophyte algae system, the application of linuron leads to a drop in the abundance of macrophytes, but an increase in the abundance of green algae. Although herbicide has direct toxicity for green algae, the nutrients released by the decomposition of macrophytes also benefited algae [SLI 05].

1.4.2. *Contamination and impact on genetic diversity*

The main genetic impacts reported in the context of contamination studies on populations include:

– increases in mutation rates;

– changes in the whole genome of genetic diversity due to demographic bottlenecks;

– changes in alleles or genotypic frequencies caused by selection triggered by contaminants;

– changes in the dispersion methods of gene fluxes that modify the genetic relationship between populations [BIC 11].

Evidence of the direct effects of pollution on genomes has been found in land organisms and in particular in the case of contamination by radionuclides [DUB 96] and aromatic polycyclic compounds [SOM 02, YAU 00].

The indirect effects of pollutants on genomes can result from a massive reduction in the size of the population resulting from mortality or even the selection of alleles or genotypes associated with "tolerance" in the contaminated zones and elimination of the majority of "sensitive" genotypes.

Only a few studies have found evidence for the existence of genetic erosion due to a massive reduction in population size in the contaminated sites. For example, Demarais *et al.* [DEM 93] concluded that the modifications in the genetic structure of freshwater fish populations (Gila de seminuda), observed after acute accidental exposure to rotenone (insecticide), were probably due to a genetic bottleneck caused by an elevated mortality rate. Murdoch and Hebert [MUR 94] put forward the same hypothesis to explain the mitochondrial genetic diversity observed in populations of cat-fish (*Ameiurus nebulosus*) taken from contaminated sites. These selective pressures can lead to changes in the distribution frequencies of alleles and genotypes in exposed populations [HEI 97, MAR 03, MUL 02, VIR 03]. Analyses of persistent fish populations surviving in chronically contaminated estuaries for more than 60 years in the United States and Europe, compared to populations living in untouched estuaries, have furnished new information on the identification of candidate genes potentially implicated in the response of the fish to chemical agents [HAH 04, HAH 05, MAR 10]. Thus, Hemmer-Hansen *et al.* [HEM 07] have demonstrated the flounder's excellent capacity to adapt (*Platichthys flesus*) to contrasting environments in estuaries. They have underlined the genetic base of these species' adaptation to the particular environment of the Baltic Sea compared to the North Sea (salinity gradient, chemical stress, hypoxia, etc.). Later on, different studies carried out on populations of flounder in estuaries confirmed these results [LAR 02, MAR 03, MAR 04]. The authors found a greater capacity to maintain the integrity of DNA among individuals carrying the allele PGM-85 in different estuaries contaminated along the Atlantic Coast of France (the estuaries of the Loire, Seine and Vilaine), compared to an uncontaminated control estuary (Ster); the hypothesis of selective pressure acting on the locus PGM has been formulated. Associations (genotype–phenotype) have been observed for several aquatic species exposed to diverse contaminants [GIL 99, KAM 00 VAN 00, WEI 02]. Correlations between individual heterozygosis and fitness components (such as survival, stability of development, growth rate, fecundity and metabolism) have been observed among different aquatic species exposed to contamination. [BEN 92, GIL 99, HAR 04, KOP 92, MAR 03, MAR 04] have observed that flounder displaying the greatest genetic variability (the most heterozygotic) were also the most capable of maintaining DNA integrity in the contaminated estuaries, confirming the conclusions of several studies carried out on natural fish populations exposed to complex mixtures of contaminants [LAR 01, LAR 02]. These correlations can in part be explained by a slower base metabolism among the heterozygotic individuals, thus leading to a smaller energy need for vital functions and finally to an improved capacity to adapt to environmental stress [DEP 96, HAU 03]. Numerous studies suggest that contamination can constitute a strong selective pressure, susceptible to leading to adaptive changes in the natural populations.

1.4.3. *Host–parasite interactions*

Parasites represent one of the most important natural biotic factors of environmental stress on natural populations. They can modify the physiology and metabolism of animals, and in consequence can influence characteristics of the lifecycle such as survival, growth and fecundity. Parasite infection can modify intra- and inter-species competition for resources, by rendering the infected individuals less competitive [REL 06].

Other stress factors, such as chemical products in the form of heavy metals or organic compounds, can exacerbate the harmful effects of infection. Conversely, a greater vulnerability of parasites to contaminants can lead to a decrease in the rate of infection [BOO 05]. Eira *et al.* [EIR 09] indicated that infestations of cestodes in eels (*Anguilla anguilla*) can modify the metabolic/storing processes of metals in the host's tissues and thus reduce the bodily charge in chrome and nickel.

Numerous chemical compounds have a tendency to alter the immune function of organisms, which is an animal's main system of protection against infections [GAL 01, GAL 03]. Two types of effect can be seen: first, the impact of pathogenic agents on animals present in the polluted zones and second, the interactions between the pesticides and biological combating agents [HOL 10]. Thus, environmental contaminants and infectious diseases are considered to be the main factors contributing to the global decline in amphibians [FOR 06, KIE 02]. Eder *et al.* [EDE 07] have found evidence of a significant infestation by the hematopoietic necrosis virus that infects salmon (*Oncorhynchus tshawytscha*) following exposure to pesticides. Several examples of evidence have also been found in fish (contamination by PCB [DUF 00] or sea-dwelling mammals). The strongest correlations between parasites and chemical contaminants have been discovered among vertebrates. However, Heinonen *et al.* [HEI 00] have shown differences in the toxicokinetic of benzo(a)pyrene between infected and healthy clams (*Pisidium amnicum*).

To protect cultures against threats in the most efficient way, conventional pesticides can be used jointly with biological combating agents. Koppenhoffer *et al.* [KOP 00] have revealed synergic interactions of imidacloprid (insecticide) applied jointly with entomopathogenic nematodes (*Steinernema glaseri* and *Heterorhabditis bacteriophora*) against white grubs (*Clidemia hirta*, *Clintonia borealis* and *Pieris japonica*). The main factor responsible for this synergic interaction between the pesticide and nematodes seems to be the general disruption of the nervous system due to the imidacloprid.

1.4.4. *Resilience and resistance*

Resistance is defined here as the capacity of a community to maintain conditions of equilibrium and to maintain its functions following exposure to contaminants. However, resilience is defined as a community's capacity to return to its initial state or to a state that enables the maintenance of its functions after exposure to toxic compounds. A better comprehension of the ecological factors that determine a system's resistance and resilience will enable observers to improve their capacity to predict how communities react to and recover from exposure to xenobiotics [CLE 02, PAL 08].

When an environmental condition – such as the atmospheric temperature and the rate at which oxygen dissolves – changes gradually over the course of time (Figure 1.8(a)), we might expect that ecosystems themselves respond gradually (Figure 1.8(b)). However, this is not always the case. Under the effect of growing pressure, certain ecosystems reach points of no return, at which they undergo sudden and unexpected transformations called "catastrophic transitions" (Figure 1.8(c)) [SCH 01].

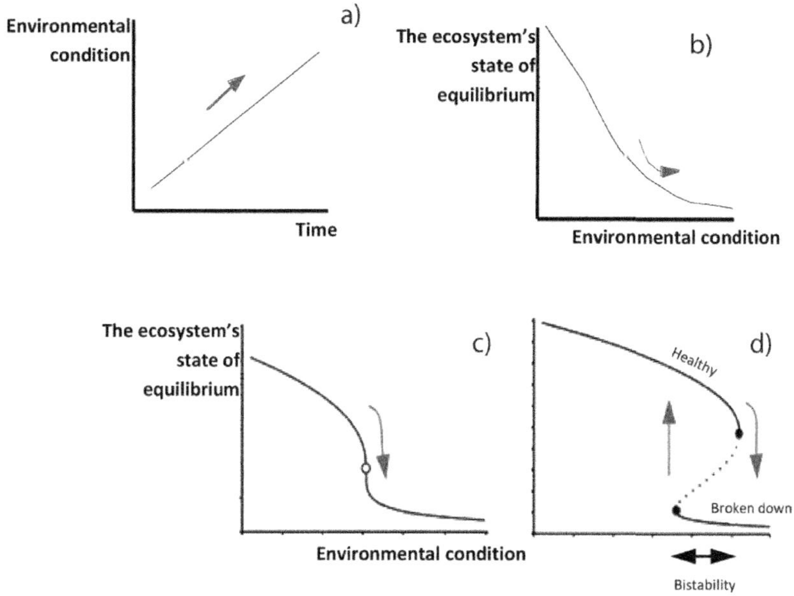

Figure 1.8. *Types of ecosystem response to a change in environmental condition (taken from [SCH 01])*

COMMENTARY ON FIGURE 1.8.– (a) Environmental condition varying gradually over time (for example, temperature, supply of nutrients or contaminants). (b)–(d): three types of ecosystem responses to these changes. (b) Continuous, gradual transition: the state of the ecosystem varies gradually in response to the change in the environmental condition. (c) Continuous, abrupt transition: the ecosystem's response becomes abrupt, and therefore less predictable but remains reversible. (d) Discontinuous transition (or catastrophic transition): the state of the system varies little until a threshold value for the environmental condition is reached. The ecosystem then swings to another state and then another mode of functioning.

Because of the ecosystems' diversity, the resilience of a community should be assessed while taking account of other stress factors such as climate change and/or invasive species [ADA 05a]. Thus, Kaufman [KAU 82] has reported that the communities associated with environments with an elevated level of stress were more tolerant than those from stable environments. The hypothesis put forward is that communities in disrupted natural habitats are preadapted to disturbances and effectively more resistant to anthropogenic stress factors [KIF 96]. However, it has been established that the effects of multiple disturbances are not necessarily cumulative [PAI 98]; consequently, the superposition of disturbances in an ecosystem already subject to chemical stress can trigger unexpected reactions from an ecological point of view [BEL 04, SCH 01].

Resistance to contaminants has been identified in numerous taxa, from microorganisms to vertebrates and vegetation [AMI 11, FEC 14] and results from two processes: (1) a physiological acclimatizing by individuals and/or (2) evolutionary adaptation in a population where the contaminants can be considered as selective factors. However, manifestations of resistance (adaptive responses) in populations exposed to complex mixes of pollutants, in fresh water or seawater, are still rare in the literature.

Individuals living in contaminated systems can produce descendants (generation F1 and F2) who display a resistance to pollutants similar to that observed for their parents; thus, this tolerance can be considered as a genetic adaptation of the population [BEL 01, JOH 11]. By using this approach, [KLE 01] have studied the Cyprinodon fish's potential resistance to chemical stress in experimental conditions. They underlined a decrease in the heritability of resistance to chemical stress with the increasing number of pollutants in the mixture. First, artificial selection carried out in a laboratory on another fish (*Heterandria formosa*) showed a rapid response to selection following exposure to cadmium; after a single generation, a proportion of two selected lines of descent out of three displayed an increase in resistance to cadmium [XIE 04]. Several studies have highlighted the fact that populations of flounder (*P. flesus*) in three chronically polluted estuary systems in the United States are resistant to aromatic hydrocarbons in their environment, compared to fish in the region of habitats little impacted by HAP. Resistance in the first generation, and sometimes in the embryos of second generation, suggests that differential survival can be due to genetic adaptation rather than to physiological acclimatization [BUR 07, VAN 08].

The energy an organism allots to developing its resistance to stress increases its probability of survival, but can have repercussions for the energy necessary for essential functions such as growth and reproduction. These energetic compromises between maintenance (survival) and production (measured, for example, by the rate of development and fertility) are found in populations subjected to chemical stress [MOU 11]. In their summary on the cost of tolerance to toxins, measured by experimental approaches, Van den Brink [VAN 00] have underlined the cost of tolerance to metals in plants, of invertebrates and fish, which show modifications in the traits of their life history. Arthropods' resistant to insecticides show a cost that is characterized by a loss of physiological performance in terms of fecundity, rates of development and fertility [ROU 87].

Several studies, carried out in laboratories, based on the successive generations of fish selected for their resistance to contamination, have typified the cost of resistance, showing:

– a reduction in the size of the offspring, a weaker fecundity and increase in the age of first maturity, for fish exposed to cadmium compared to control fish [XIE 04];

– a greater vulnerability in fish exposed to contaminated estuary sediments compared to control fish faced with other stress factors such as ultraviolet (UV) radiation and hypoxia [MEY 03];

– in the case of flounder (*P. flesus*), resistant genotypes, characterized by their significant capacity to maintain the integrity of their DNA in polluted estuaries, have shown a reduced fecundity and a weaker condition index [MAR 04].

Different molecular mechanisms for resisting toxins, therefore, generate energy costs that can affect essential physiological functions [TAY 96]. However, the majority of studies carried out in this domain were performed with experimental measurement in laboratory, and a major question remains: what is the validity of costs estimated in a laboratory, in a natural environment?

A system's "recuperation" time (resilience) depends on the capacity of species to recolonize the system (generation time) [BRO 00] and the half-life of the contaminant, which can vary from a few days to several years. The resilience can occur due to an increase in the abundance of species still in the system or by the migration of species coming from the exterior of the system. For example, studies in lotic systems have shown that the diversity of macroinvertebrates can fluctuate following the application of insecticides, because individuals can colonize sites upstream that are not affected by the insecticide [WAL 96]. In the same way, insects can recolonize lentic systems after exposure to pesticides, which contributes to the system's resilience [WAN 96]. This suggests that the geographic distribution of species and the structuration of metapopulations can play an important role in the recuperation of ecosystems.

The contaminants can have long-term effects on the ecosystems; Woin [WOI 98] has demonstrated that two years after the application of fenvalerate insecticide (a pyrethrinoid, organochlorinated insecticide), a community of invertebrates in a pond was significantly different in the diversity of the species and in abundance compared to control sites. In the same way, the application of herbicides can lead to a reduction in the biomass of producers, which persists because of recuperation periods specific to the species [SPA 97]. Overall, these studies show that the resilience of ecosystems to exposure to pesticides will depend on a certain number of parameters (beyond lethal and sublethal effects on each member of the community) such as the type of habitat, rates of migration, the dynamics of extinction and recolonization, the specific sensitivity of the species and the species' specific rates of recuperation.

The resilience *stricto sensu* of an ecosystem, following an improvement in the quality of the water and/or sediment, can be retarded by the installation of tolerant species that check recolonization by sensitive species, such as has been reported in polluted lakes [FRO 06]. The ecosystem has, therefore, evolved in a lasting manner, eventually returning to its initial functions, supported by other species.

Theoretical and empirical studies suggest that certain communities show abrupt and nonlinear changes in their structure or functioning in response to disturbances (Figure 1.8(d)) [CON 83, EST 95, MAY 77]. The concept of an ecological threshold has a direct relationship with the concepts of resistance and resilience. The threshold for resistance represents the concentrations of the contaminant that will lead to sudden modifications of the community's structure. The threshold of resilience represents "the point in time" where recuperation is initiated and finally completed after the elimination of the stress factor. This notion of a threshold is illustrated in several studies carried out on different systems such as lakes, coral reefs and the pelagic system [GRO 06, SCH 01].

1.5. Indirect effects and multiple stress factors

Generally speaking, marine ecosystems are simultaneously exposed to a combination of stress factors (contaminants, temperature, eutrophization, anoxia, etc.). Each stress factor can have an impact on an individual or the community, and their combination may or may not produce other additive effects [CAS 98]. Few studies have been carried out on the effects of multiple stress factors on biological communities [BRE 99], and even fewer have examined the consequences in terms of indirect effects [CUP 02]. Resistance to different types of contaminant is not necessarily correlated, which renders the impact of concomitant stresses unpredictable at the level of the community. For example, several studies [BRO 95, CUP 95, VAN 95] suggest that insecticides increase the negative effects of eutrophization by a "top-down" effect where the biomass of periphyton is no longer controlled by grazers vulnerable to insecticide.

One of the challenges of ecotoxicology – ecology under toxic stress – is to understand the toxic chemical pressure, amid the complex mix of multiple disruptions to natural environments. In this context, the importance of global climate change (GCC) and its potential interaction with contaminants in the environment has recently received increasing attention [CLE 09, NOY 09, SCH 07, VAN 08]. Projections from the intergovernmental group of experts on the climate *Groupe d'experts Intergouvernemental sur l'Evolution du*

Climate (GIEC), on the subject of changes for the end of this century, concern a wide range of environmental conditions: raised temperature, changes in precipitation regimes, an increase in the acidity of the oceans and a reduction in ice covering the seas [INT 07] (see also the [MON 14a, b and c] also from the Seas and Oceans set). These projections include significant uncertainties in regional variations but agree on an increase in the frequency of extreme meteorological phenomena such as heat waves, droughts and storms. The transfer, the arrival and then the exposure to contaminants will also be impacted by GCC [VER 08], although there is already great uncertainty associated with the effects of GCC on future concentrations of contaminants in the environment [NOY 09].

1.5.1. *Impact on the future of contaminants*

It is at the level of the highest latitudes that the increase in temperature is the most pronounced. Reports from IPCC indicate that the average of temperatures in the Arctic has increased by almost twice more than the global average in the last hundred years [IPC 07]. POPs are, mostly, semi-volatile compounds that can be found thousands of kilometers from the place where they were emitted. In effect, they are transported, in the form of gas and/or adsorbed to aerosols, by global atmospheric circulation into higher latitudes. As the gradients in temperature between high and low latitudes will be less pronounced, the presence of persistent organic pollutants in high latitudes could diminish [BEY 03, BRA 05, BRE 04, WAN 96]. Moreover, this increase in temperature will also have an impact on the degradation processes of POP, which are more significant at higher temperatures. It is, however, important to note that this degradation can lead to metabolites that are more toxic than the initial compounds (see [DAL 07] on PCB and the dioxins in the Venice laguna). Organic carbon cycles in land and aquatic systems will also be modified by climate change, which will have a direct impact on the bioavailability of a large number of contaminants [MAC 03, MAG 97, SCH 97]. This phenomenon has been found in the boreal lakes of north-west Ontario following a long period of warming and drought associated with forest fires [SCH 97]. Moreover, the warming will “release” more easily the contaminants “trapped” in the permafrost, rendering them “available” for exchanges with air and/or running water [MAC 05]. This availability will also augment the degradation processes [BEY 03, BRU 98, MA 04, SCH 05].

In addition to numerous abiotic factors that can influence the behavior of contaminants, modifications in the migratory habits of several species due to climate change could be an important modulating factor for the transport of POP [BLA 07]. In effect, migratory species, such as fish, birds and sea mammals, can be exposed to contaminants in an impacted geographic zone (coastal or estuary zone) and by migrating, transport these contaminants in substantial quantities to other sites that are not directly impacted. This "biotic" transport of contaminants can have an order of size similar to atmospheric and oceanic transport [BUR 08]. In effect, Blais *et al.* [BLA 07] have shown that birds from the Arctic and Antarctica act as vectors for persistent contaminants from oceans to land systems *via* their guano. Thus, on the Canadian coast under the cliffs where northern fulmars nest (*Fulmarus glacialis*), concentrations in HexaChloroBenzene (HCB), DDT and mercury are 10–60 times higher than concentrations of contaminants in sediments not impacted by nesting. Similar results have been observed for Antarctica [BLA 07]. Thus, if climate change modifies the migrations of several species, the local and global transport of POPs will also be modified with impacts on other ecosystems [BUR 08, WRO 05].

Finally, the adaptation of societies to climate change (CC) also risks affecting the presence of contaminants in different regions due to modifications in agricultural practices, the more significant use of pesticides due to the proliferation of pests [KAT 11] and an increase in the exploitation of resources from polar regions [DE 11].

1.5.2. *Effects of contaminants and climate change on different organization levels of life forms*

Physiological mechanisms implicated in the combined effects of toxic substances and climatic stresses can be interpreted from two different angles, as proposed by [HOO 12] (Figure 1.9): (1) the increased sensitivity to toxins caused by the climate, where exposure to a stress factor linked to the climate renders an organism more sensitive to exposure to toxic substances (Figure 1.9, arrow 3), and (2) exposure to a contaminant renders an organism more vulnerable to climate changes (Figure 1.9, arrow 4).

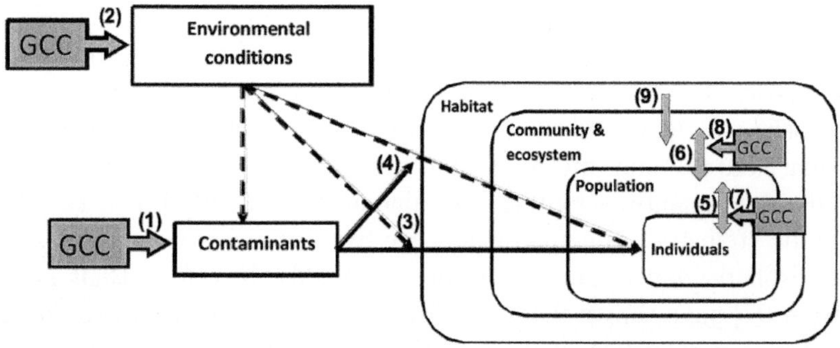

Figure 1.9. *Combined effects of the impact of global climate change (GCC) and contaminants on different levels of biological organization (taken from [MOE 13])*

COMMENTARY ON FIGURE 1.9.– The term "GCC" represents the climatic factors, such as the temperature and precipitation. The environmental conditions represent other abiotic factors (for example, hydrological regimes, UV radiation and concentrations of nutrients). GCC can affect the transfer and exposure of toxic products directly (arrow 1) or by means of environmental conditions (arrow 2) [GOU 13]. Individuals can be affected by GCC due to exposure to toxic substances and/or other environmental conditions; interactions between these factors can lead to a sensitivity to toxins caused by the climate (arrow 3) or a sensitivity to climate change caused by the contaminant (arrow 4) [HOO 12]. The combined effects of the contaminants and GCC on the individuals can spread to higher levels of biological organization (arrows 5 and 6). These transfer processes in turn can be affected by GCC, directly or indirectly (arrows 7 and 8). Finally, the properties habitats can influence the responses of populations and communities to the combined effects of contaminants and GCC (arrow 9).

Interactions between different environmental stresses can be synergic, antagonistic or cumulative. They vary depending on the trophic level (herbivore and predator) and the level of response (individual, population and community) [CRA 08]. It is, therefore, difficult to predict the impact that the effect of the climate, combined with chemical stress observed at an individual level, will have on other levels of biological organization [DE 11, PAI 98].

Many authors have found evidence that climate change has affected the phenology of organisms and the geographical distribution of species, as well as the composition and dynamic of communities [LOV 05, PEN 01, ROO 03, WAL 01].

The toxico-kinetic accumulation and the toxicity of POPs and pesticides in ecosystems are susceptible to increasing in response to the increase in temperatures and salinity [CAP 06, GAU 00, HEU 01, MOO 03, SCH 07, WAN 01, WAR 04]. A study carried out by Maruya *et al.* [MAR 05] on an estuary-dwelling fish (*F. heteroclitus*) has thus found evidence that rates of elimination of several toxaphene congeners are higher at 25°C than in water at 15°C. Similar results have been observed among perch on elimination rates of different PCB congeners over a seasonal cycle [PAT 07a] and on rainbow trout when following hydroxylation rates for PCB [BUC 07].

In parallel with these processes, the increase in temperature can also alter key physiological functions, aggravating the harmful effects of contaminants [BRO 02, BRO 04]. Although the exact mechanisms that underlie this relationship are not fully understood, several studies indicate that the temperature causes changes in the metabolism and notably in the activation of enzymatic systems implicated in the detoxification processes [BUC 07, LYD 99]. Thus, Capkin *et al.* [CAP 06] have shown that mortality rates in rainbow trout (*Oncorhynchus mykiss*) exposed to the insecticide endosulfan were higher at 16°C than at 13°C. Monserrat and Bianchini [MON 95] suggest a similar explanation for the increase in the toxicity of methyl parathion for crabs (*Chasmagnathus granulata*) after exposure to temperatures of 12–30°C. Conversely, DDT is generally considered to be more toxic at low temperatures, which could be due to an increase in the modulation sodium channels, which would lead to vulnerability in the nervous system [NAR 00].

Species living at the limits of their homeostatic or physiological tolerance will then be the most vulnerable to the double stress caused by climate change and exposure to contaminants [GOR 03, HEA 94, PAT 07]. In effect, the capacity of species and populations to tolerate raised temperatures can be altered by toxic products. This double exposure acts as a "co-stressor": the toxic substances affecting the physiological functions can diminish the capacity of organisms to maintain homeostasy [BRO 04]. Ectoderms, like fish, are particularly vulnerable to these temperature–contaminant interactions. The capacity of brook trout (*Salvelinus fontinalis*) and Atlantic salmon (*Salmo salar*) to acclimatize to an increase in temperature is weakened by sublethal doses of DDT [AND 69].

Moreover, modifications in habitat and trophic networks caused by climate change can affect the toxicity of contaminants by modifying the means of exposure and the sensitivity of certain populations, in particular those that are already in a stress situation [BRE 04, BRO 02, GAS 03, GIL 03, MAC 05].

Finally, these interactions between climate change and exposure to contaminants will also be dependent on the life stage subject to this exposure. In effect, the sensitive stages of life induce responses that, in their turn, modify the physiological processes. This process is particularly marked in contamination by endocrine disruptors [BRI 08].

1.6. Conclusion

Marine ecotoxicology is a very recent avenue of research. Until the 1990s, the dilution capacity of the oceans appeared to be so great that a "drop of PCB in the sea" would not have any consequences... Indeed, it is no longer possible to ignore the harmful impact of xenobiotic contaminants. First, it has been proved that they are present in the confines of the ocean, in the environment as much as in the fauna to which it is home. Second, the chronic toxicity that they cause constitutes a still unvalidated but probable hypothesis for explaining certain collapses in marine species that are not otherwise understood: such as stocks of farmed species not being replaced after the cessation of farming, non-farmed species in decline and immune deficiencies in cultivated species... Finally, the contamination of sea produce consumed by mankind poses proven health problems and triggers numerous halts in fishing activity (eels, shad, sardines, etc.).

Even more than for aquatic continental ecosystems, the object of marine ecotoxicology is necessarily ecology under chronic toxic stress. In effect, the contaminants involved are transported over tens of thousands of kilometers, in a relatively little-fragmented environment. They can survive over decades and infect the whole of the trophic chain. These scales of time and biological organization immediately pose questions at the level of ecology and indirect effects.

However, knowledge of them remains limited, since relatively few experiments have been devised specifically to test it. Changes in behavior, physiology, trophic interactions and/or competition between species can produce changes in populations and the composition of the community that intensify or mask the direct toxic effects. "Trophic cascades" seem to be a shared type of indirect effect, but the setup of most of the experiments does not permit the univocal distinction between trophic interactions and interactions involving competition; these two types of response are indirect effects, but result from very different ecological processes.

More work is also necessary to understand the multiple stress factors and know how, directly and indirectly, they influence the structure and behavior of aquatic communities. Similarly, complementary research should be carried out to develop ecosystem models that describe and predict direct and indirect effects of contaminants on a large variety of aquatic habitats. The aim of these approaches should be to understand the ecological implications of environmental stress factors, and, finally, to aid the development of management strategies for preserving and restoring the integrity of natural habitats. To understand the direct toxic effects of the contaminants is, and will continue to be, an important part of this process. However, communities and ecosystems are much more than the sum of their components, and the challenge to come is to understand the sometimes subtle but important integration of the indirect influence of contaminants in a realistic approach to exposure.

Experiments in micro- and meso-cosmes constitute necessary tools for studying the combined effects of different environmental stresses; although there is disagreement and debate concerning the modeling methods for the analysis and interpretation of these data [LIE 11, VAN 12]. Long-term ecotoxicological experiments that integrate the combination of climate evolution and environmental variations on a credible scale would permit more reliable predictions of the impact of toxic products in the context of climate change. Three approaches seem particularly promising:

– studying the potential of species to adapt, with the help of genetic analyses (variability and correlations);

– to carry out experiments (in a controlled setting) on microevolution;

– to compare populations tolerant to climate changes with populations resistant to toxic substances [SCH 11].

Marine ecosystems have been living under toxic pressure, probably for decades, but the level of this pressure has increased considerably with the rapid expansion of synthetic chemistry and the production of manufactured goods from the middle of the 20th Century. If marine life has developed with ingenious mechanisms enabling it to regulate, indeed to use, metals that are a constituent of the Earth's crust, it is not necessarily the same for the new, synthetic organic molecules, which are endlessly renewed, with which it is faced today. The formidable distances over which these substances are transported, physically or biologically, separate the origins and impacts of these pollutions in space and time. Everything should be done to reduce contamination on such a vast scale in environments that have a heritage value and are rich in practical uses.

But, we also have a lot to learn about the dynamics of these marine ecosystems that have developed strategies of maintenance, indeed of development. Life is fundamentally fluid, "unpredictable and uncertain [...]. We should credit it with the ability to surprise us" advises Tassin [TAS 14]. To study ecological mechanisms developed in the face of toxic pressure without let up can also give us keys to understanding the living world that may help the human species itself to live with global changes, hand-in-hand with the trajectory of the biosphere of which it is part.

1.7. Bibliography

[ABI 12] ABIDLI S., MACHADO SANTOS M., LAHBIB Y., *et al.*, "Tributyltin (TBT) effects on *Hexaplex trunculus* and *Bolinus brandaris* (Gastropoda:Muricidae): imposex induction and sex hormone levels insights", *Ecological Indicators*, vol. 13, pp. 13–21, 2012.

[ADA 00] ADAMS M., GREELEY M.S., "Ecotoxicological indicators of water quality: using multi-response indicators to assess the health of aquatic ecosystems", *Water, Air, & Soil Pollution*, vol. 123, pp. 103–115, 2000.

[ADA 05a] ADAMS M., "Assessing cause and effect of multiple stressors on marine systems", *Marine Pollution Bulletin*, vol. 51, pp. 649–657, 2005.

[ADA 05b] ADAMS M., RYON M.G., SMITH J.G., "Recovery in diversity of fish and invertebrate communities following remediation of a polluted stream: investigating causal relationships", *Hydrobiologia*, vol. 542, pp. 77–93, 2005.

[AMI 11] AMIARD-TRIQUET C., "Pollution tolerance: from fundamental biological mechanisms to ecological consequences", in AMIARD-TRIQUET C., RAINBOW P.S., ROMÉO M. (eds.), *Tolerance to Environmental Contaminants*, CRC Press, Boca Raton, pp. 1–24, 2011.

[AND 69] ANDERSON J.M., PETERSON M.R., "DDT – sublethal effects on brook trout nervous system", *Science*, vol. 164, no. 3878, pp. 440–441, 1969.

[AND 92] ANDERSSON T., FÖRLIN L., "Regulation of the cytochrome P450 enzyme system in fish", *Aquatic Toxicology*, vol. 24, pp. 1–20, 1992.

[BAZ 97] BAZZANTI M., CHIAVARINI S., CREMISINI C., *et al.*, "Distribution of PCB congeners in aquatic ecosystems: a case study", *Environment International*, vol. 23, pp. 799–813, 1997.

[BEL 00] BELANGER S.E., GUCKERT J.B., BOWLING J.W., *et al.*, "Responses of aquatic communities to 25-6 alcohol ethoxylate in model stream ecosystems", *Aquatic Toxicology*, vol. 48, pp. 135–150, 2000.

[BEL 01] BELFIORE N.M., ANDERSON S.L., "Effects of contaminants on genetic patterns in aquatic organisms: a review", *Mutation Research*, vol. 489, pp. 97–122, 2001.

[BEL 04] BELLWOOD D.R., HUGHES T.P., FOLKE C., *et al.*, "Confronting the coral reef crisis", *Nature*, vol. 429, pp. 827–833, 2004.

[BEN 92] BENTON M.J., GUTTMAN S.I., "Allozyme genotype and differential resistance to mercury pollution in the caddisfly, *Nectopsyche albida*: II. Mutilocus genotypes", *Canadian Journal of Fisheries and Aquatic Sciences.*, vol. 49, pp. 147–149, 1992.

[BEY 03] BEYER A., WANIA F., GOUIN T., *et al.*, "Temperature dependence of the characteristic travel distance", *Environmental Science & Technology*, vol. 37, pp. 766–771, 2003.

[BIC 11] BICKHAM J.W., "The four cornerstones of evolutionary toxicology", *Ecotoxicology*, vol. 20, pp. 497–502, 2011.

[BLA 01] BLACK W.C., BAER C.F, ANTOLIN M.F, *et al.*, Population genomics: genome-wide sampling of insect populations", *Annual Review of Entomology*, vol. 46, pp. 441–469, 2001.

[BLA 07] BLAIS J.M., MACDONALD R.W., MACKEY D., *et al.*, "Biologically mediated transport of contaminants to aquatic systems", *Environmental Science & Technology*, vol. 41, no. 4, pp. 1075–1084, 2007.

[BOG 96] BOGOMOLOV D.M., CHEN S.K., PARMELEE R.W., *et al.*, "An ecosystem approach to soil toxicity testing: a study of copper contamination in laboratory soil microorganisms", *Applied Soil Ecology*, vol. 4, pp. 95–105, 1996.

[BOO 05] BOONSTRA R., MANZON R.G., MIHOK S., *et al.*, "Hormetic effects of gamma radiation on the stress axis of natural populations of meadow voles (Microtus pennsylvanicus)", *Environmental Toxicology and Chemistry*, vol. 24, pp. 334–343, 2005.

[BRA 05] BRAUNE B.M., OUTRIDGE P.M., FISK A.T., *et al.*, "Persistent organic pollutants and mercury in marine biota of the Canadian Arctic: an overview of spatial and temporal trends", *Science of the Total Environment*, vol. 351, pp. 4–56, 2005.

[BRE 99] BREITBURG D.L., SANDERS J.G., GILMOUR C.C., *et al.*, "Variability in responses to nutrients and trace elements, and transmission of stressor effects through an estuarine food web", *Limnology and Oceanography*, vol. 44, pp. 837–863, 1999.

[BRE 04] BREIVIK K., ALCOCK R., LI Y.F., *et al.*, "Primary sources of selected POPs: regional and global scale emission inventories", *Environmental Pollution*, vol. 128, nos. 1–2, pp. 3–16, 2004.

[BRI 08] BRIAN J.V., HARRIS C.A., RUNNALLS T.J., *et al.*, "Evidence of temperature-dependent effects on the estrogenic response of fish: implications with regard to climate change", *Science of the Total Environment*, vol. 397, pp. 72–81, 2008.

[BRO 82] BROECLKER W., PENG T., *Tracers in the Sea*, Eldigio, Palisades, p. 691, 1982.

[BRO 95] BROCK T.C.M., ROIJACKERS R.M.M., ROLLON R., *et al.*, "Effects of nutrient loading and insecticide application on the ecology of *Elodea*-dominated freshwater microcosms II. Responses of macrophytes, periphyton and macroinvertebrate grazers", *Archiv fur Hydrobiologie*, vol. 134, pp. 53–74, 1995.

[BRO 00] BROCK T., LAHR J., VAN DEN BRINK P.J., *Ecological Risks of Pesticides in Freshwater Ecosystems. Part 1: Herbicides*, Alterra Green World Research, Alterra, Wageningen, Netherlands, 2000.

[BRO 02a] BROOK R.K., RICHARDSON E.S., "Observations of polar bear predatory behaviour toward caribou", *Arctic*, vol. 55, no. 2, pp. 193–196, 2002.

[BRO 02b] BROOMHALL S., "The effects of endosulfan and variable water temperature on survivorship and subsequent vulnerability to predation in *Litoria citropa* tadpoles", *Aquatic Toxicology*, vol. 61, nos. 3–4, pp. 243–250, 2002.

[BRO 04] BROOMHALL S.D., "Egg temperature modifies predator avoidance and the effects of the insecticide endosulfan on tadpoles of an Australian frog", *Journal of Applied Ecology*, vol. 41, no. 1, pp. 105–113, 2004.

[BRO 10] BRODMAN R., NEWMAN W.D., LAURIE K., *et al.*, "Interaction of an aquatic herbicide and predatory Salamander density on wetland communities", *Journal of Herpetology*, vol. 44, pp. 69–82, 2010.

[BRU 98] BRUBAKER W.W., HITES R.A., "OH reaction kinetics of polycyclic aromatic hydrocarbons and polychlorinated dibenzo-p-dioxins and dibenzofurans", *Journal of Physical Chemistry A*, vol. 102, no. 6, pp. 915–921, 1998.

[BRU 01] BRUNO J.F., BERTNESS M.D., "Habitat modification and facilitation in benthic marine communities", in BERTNESS M.D., GAINES S.D., HAY M.E. (eds.), *Marine Community Ecology*, Sinauer Associates, Sunderland, pp. 201–218, 2001.

[BUC 07] BUCKMAN A.H., BROWN S.B., SMALL J., *et al.*, "Role of temperature and enzyme induction in the biotransformation of polychlorinated biphenyls and bioformation of hydroxylated polychlorinated biphenyls by rainbow trout (Oncorhynchus mykiss)", *Environmental Science and Technology*, vol. 41, pp. 3856–3863, 2007.

[BUR 94] BURGEOT T., BOCQUENE G., PINGRAY G., *et al.*, "Monitoring biological effects of contamination in marine fish along French coast by measurement of ethoxyresorufin-o-deethylase activity", *Ecotoxicology and Environmental Safety*, vol. 29, pp. 131–147, 1994.

[BUR 07] BURNETT K.G., BAIN LISA J., BALDWIN WILLIAM S., *et al.*, "*Fundulus* as the premier teleost model in environmental biology: opportunities for new insights using genomics", *Comparative Biochemistry and Physiology*, vol. 2D, pp. 257–86, 2007.

[BUR 08] BUREK K.A., GULLAND F.M.D., OHARA T.M., "Effects of climate change on Arctic marine mammal health", *Ecological Applications*, vol. 18, no. 2, pp. S126–S134, 2008.

[CAI 83] CAIRNS J., "Are single species toxicity tests alone adequate for estimating environmental hazard?", *Hydrobiologia*, vol. 100, pp. 47–57, 1983.

[CAJ 03] CAJARAVILLE M.P., HAUSER L., CARVALHO G., *et al.*, "Genetic damage and the molecular/cellular response to pollution", in LAWRENCE A.J., HEMINGWAY K.L. (eds.), *Effects of Pollution on Fish, Molecular Effects and Population Responses*, Blackwell Sciences, Oxford, pp. 14–82, 2003.

[CAL 83] CALABRESE E., *Principles of Animals Extrapolations*, John Wiley & Sons, New York, NY, 1983.

[CAL 97] CALOW P., SIBLY R., FORBES V., "Risk assessment on the basis of simplified life history scenarios", *Envionmental Toxicology and Chemistry*, vol. 16, pp. 1983–1989, 1997.

[CAN 02] CANIVET V., GIBERT J., "Sensitivity of epigeanand hypogean freshwater macroinvertebrates to complex mixture. Part I: laboratory experiments", *Chemosphere*, vol. 46, pp. 999–1009, 2002.

[CAN 10] CANESI L., FABBRI R., GALLO G., *et al.*, "Biomarkers in Mytilus galloprovincialis exposed to suspensions of selected nanoparticles (nano carbon black, C60 fullerene, nano-TiO2, nano-SiO2)", *Aquatic Toxicology*, vol. 100, no. 2, pp. 168–177, 2010.

[CAP 06] CAPKIN E., ALTINOK I., KARAHAN S., "Water quality and fish size affect toxicity of endosulfan, an organochlorine pesticide, to rainbow trout", *Chemosphere*, vol. 64, no. 10, pp. 1793–800, 2006.

[CAR 97] CARMAN K.R., FLEEGER J.W., POMARICO S., "Response of a benthic food web to hydrocarbon contamination", *Limnology and Oceanography*, vol. 42, pp. 561–571, 1997.

[CAS 98] CASSEE F.R., GROTEN J.P., VAN BLADEREN P.J., *et al.*, "Toxicological evaluation and risk assessment of chemical mixtures", *Critical Reviews in Toxicology*, vol. 28, pp. 73–101, 1998.

[CAS 01] CASWELL H., *Matrix Population Models: Construction, Analysis and Interpretation*, 2nd ed., Sinauer Associates, Sunderland, 2001.

[CAS 06] CASELLAS C., "Risques écologiques et sanitaires des résidus de médicaments dans les eaux", *Colloque Hydrosciences*, Montpellier, France, 16–17 June 2005, proceedings published in *Environnement, Risques & Santé* (ERS), vol. 5, no. 4, p. 225, 2006.

[CLE 94] CLEMENTS W.H., KIFFNEY P.M., "Assessing contaminant effects at higher levels of biological organization", *Environmental Toxicology and Chemistry*, vol. 13, pp. 357–359, 1994.

[CLE 99] CLEMENTS W.H., "Metal tolerance and predator-prey interactions in benthic macroinvertebrate stream communities", *Ecological Applications*, vol. 9, pp. 1073–1084, 1999.

[CLE 02] CLEMENTS W.H., NEWMAN M.C., *Community Ecotoxicology*, John Wiley & Sons, 2002.

[CLE 09] CLEMENTS W.H., ROHR J.R., "Community responses to contaminants: using basic ecological principles to predict ecotoxicological effects", *Environmental Toxicology and Chemistry*, vol. 28, pp. 1789–1800, 2009.

[CON 83] CONNELL J.H., SOUSA W.P., "On the evidence needed to judge ecological stability or persistence", *The American Naturalist*, vol. 121, pp. 789–824, 1983.

[COS 97] COSTANZA R., DARGE R., DEGROOT R., *et al.*, "The value of the world's ecosystem services and natural capita", *Nature*, vol. 387, p. 253, 1997.

[COS 11] COSSA D., HEIMBURGER L.E., LANNUZEL D., *et al.*, "Mercury in the Southern ocean", *Geochimica et Cosmochimica Acta*, vol. 75, no. 14, pp. 4037–4052, 2011.

[CRA 08a] CRAIN D.A., JANSSEN S.J., EDWARDS T.M., *et al.*, "Female reproductive disorders: the roles of endocrine-disrupting compounds and developmental timing", *Fertility and Sterility*, vol. 90, pp. 911–940, 2008.

[CRA 08b] CRAIN C.M, KROEKER K., HALPERN B.S., "Interactive and cumulative effects of multiple human stressors in marine systems", *Ecology Letters*, vol. 11, pp. 1304–1315, 2008.

[CUP 95] CUPPEN J.G.M., GYLSTRA R., VAN BEUSEKOM S., *et al.*, "Effects of nutrient loading and insecticide application on the ecology of Elodea-dominated freshwater microcosms III. Responses of macroinvertebrate detritivores, breakdown of plant litter, and final conclusions", *Archiv fur Hydrobiologie*, vol. 134, pp. 157–177, 1995.

[CUP 02] CUPPEN J.G.M., CRUM S.J.H., VAN DEN HEUVEL H., *et al.*, "Effects of a mixture of two insecticides in freshwater microcosms: I. Fate of chlorpyrifos and lindane and responses of macroinvertebrates", *Ecotoxicology*, vol. 11, pp. 165–180, 2002.

[DAL 07] DALLA VALLE M., CODATO E., MARCOMINI A., "Climate change influence on POPs distribution and fate: a case study", *Chemosphere*, vol. 67, no. 7, pp. 1287–1295, 2007.

[DAL 13] DALLAIRE R., DEWAILLY E., AYOTTE P., *et al.*, "Exposure to organochlorines and mercury through fish and marine mammal consumption: associations with growth and duration of gestation among Inuit newborns", *Environment International*, vol. 54, pp. 85–91, 2013.

[DAO 11] DAOUK T., LARCHER T., ROUPSARD F., *et al.*, "Long-term food-exposure of zebrafish to PCB mixtures mimicking some environmental situations induces ovary pathology and impairs reproduction ability", *Aquatic Toxicology*, vol. 105, pp. 270–278, 2011.

[DAS 14] DASTOOR A.P., DURNFORD D.A., "Arctic ocean: is it a sink or a source of atmospheric mercury?", *Environmental Science & Technology*, vol. 48, no. 3, pp. 1707–1717, 2014.

[DE 89] DE NOYELLES F. JR., KETTLE W.D., FROMM C.H., *et al.*, "Use of experimental ponds to assess the effects of a pesticide on the aquatic environment", *Miscellaneous Publications Entomological Society of America*, vol. 75, pp. 41–56, 1989.

[DE 09] DE MARSILY G., *L'eau, un trésor en partage*, Dunod, Paris, 2009.

[DE 11] DE HOOP L., SCHIPPER A.M., LEUVEN R.S.E.W., *et al.*, "Sensitivity of polar and temperate marine organisms to oil components", *Environmental Science & Technology*, vol. 45, pp. 9017–9023, 2011.

[DED 09] DEDOURGE-GEFFARD O., PALAIS F., BIAGIANTI-RISBOURG S., *et al.*, "Effects of metals on feeding rate and digestive enzymes in Gammarus fossarum: an *in situ* experiment", *Chemosphere*, vol. 77, no. 11, pp. 1569–1576, 2009.

[DEM 93] DEMARAIS B.D., DOWLING T.E., MONCKLEY W.L., "Post-perturbation genetic changes in populations of endangered Virgin River chubs", *Conservation Biology*, vol. 7, pp. 334–41, 1993.

[DEP 96] DEPLEDGE M.H., "Genetic ecotoxicology: an overview", *Journal of Experimental Marine Biology and Ecology*, vol. 200, pp. 57–66, 1996.

[DEW 08] DEWAILLY E., SUHAS E., MOU Y., *et al.*, "High fish consumption in French Polynesia and prenatal exposure to metals and nutrients", *Asia Pacific Journal of Clinical Nutrition*, vol. 17, no. 3, pp. 461–470, 2008.

[DIE 13] DIETZ R., SONNE C., BASU N., *et al.*, "What are the toxicological effects of mercury in Arctic biota?", *Science of the Total Environment*, vol. 443, pp. 775–790, 2013.

[DOD 95] DODSON S.I., HANAZATO T., GORSKI P.R., "Behavioral responses of *Daphnia pulex* exposed to carbaryl and *Chaoborus* Kairomone", *Environmental Toxicology and Chemistry*, vol. 14, pp. 43–50, 1995.

[DUB 96] DUBROVA Y.E., NESTEROV V.N., KROUCHINSKY N.G., *et al.*, "Human minisatellite mutation rate after the Chernobyl accident", *Nature*, vol. 380, pp. 683–686, 1996.

[DUF 00] DUFFY J.E., HAY M.E., "Strong impacts of grazing amphipods on the organization of a benthic community", *Ecological Monographs*, vol. 70, pp. 237–263, 2000.

[EDE 07] EDER K.J., KÖHLER H.R., WERNER I., "Pesticide and pathogen: heat shock protein expression and acetylcholinesterase inhibition in juvenile Chinook salmon in response to multiple stressors", *Environmental Toxicology and Chemistry*, vol. 26, pp. 1233–1242, 2007.

[EIR 09] EIRA C., TORRES J., MIQUEL J., *et al.*, "Trace element concentrations in *Proteocephalus macrocephalus* (Cestoda) and *Anguillicola crassus* (Nematoda) in comparison to their fish host, *Anguilla anguilla* in Ria de Aveiro, Portugal", *Science of the Total Environment*, vol. 407, pp. 991–998, 2009.

[ESC 02] ESCHER B., HERMENS J., "Modes of action in ecotoxicologie: their role in body burdens, species sensitivity, QSARs, and mixture effects", *Environmental Sciences and Technology*, vol. 36, pp. 4201–4217, 2002.

[EST 95] ESTES J.A., DUGGINS D.O., "Sea otters and kelp forests in Alaska – generality and variation in a community ecological paradigm", *Ecological Monographs*, vol. 65, pp. 75–100, 1995.

[EVE 94] EVERAARTS J.M., SLEIDERINK H.M., DEN BESTEN P.J., *et al.*, "Molecular responses as indicators of marine pollution: DNA damage and enzyme induction in *Limanda limanda* and *Asterias rubens*", *Environmental Health Perspectives*, vol. 102, Suppl. 12, pp. 37–43, 1994.

[FEC 14] FECHNER L., GOURLAY-FRANCÉ C., TUSSEAU-VUILLEMIN M.H., "Linking community tolerance and structure with low metallic contamination: a field study on 13 biofilms sampled across the Seine river basin", *Water Research*, vol. 51, pp. 152–162, 2014.

[FLE 03] FLEEGER J.W., CARMAN K.R., NISBET R.M., "Indirect effects of contaminants in aquatic ecosystems", *Science of the Total Environment*, vol. 317, pp. 207–233, 2003.

[FOR 06] FORSON D., STORFER A., "Effects of atrazine and iridovirus infection on survival and lifehistory traits of the long-toed salamander (*Ambystoma macrodactylum*)", *Environmental Toxicology and Chemistry*, vol. 25, pp. 168–173, 2006.

[FRO 06] FROST T.M., FISCHER J.M., KLUG J.L., *et al.*, "Trajectories of zooplankton recovery in the Little Rock Lake whole-lake acidification experiment", *Ecological Applications*, vol. 16, pp. 353–367, 2006.

[GAL 91] GALGANI F., BOCQUENE G., LUCON M., *et al.*, "EROD measurements in Fis from the Northwest Part of France", *Marine Pollution Bulletin*, vol. 22, pp. 494–500, 1991.

[GAL 01] GALLOWAY T.S., DEPLEDGE M.H., "Immunotoxicity in invertebrates: measurement and ecotoxicological relevance", *Ecotoxicology*, vol. 10, pp. 5–23, 2001.

[GAL 03] GALLOWAY T., HANDY R., "Immunotoxicity of organophosphorous pesticides", *Ecotoxicology*, vol. 12, pp. 345–363, 2003.

[GAS 03] GASTON A.J., WOO K., HIPFNER J.M., "Trends in forage fish populations in northern Hudson Bay since 1981, as determined from the diet of nestling thick-billed murres *Uria lomvia*", *Arctic*, vol. 56, no. 3, pp. 227–233, 2003.

[GAU 00] GAUNT P., BARKER S.A., "Matrix solid phase dispersion extraction of triazines from catfish tissues; examination of the effects of temperature and dissolved oxygen on the toxicity of atrazine", *International Journal of Environment and Pollution*, vol. 13, nos. 1–6, pp. 284–312, 2000.

[GEO 94] GEORGE S.G., "Enzymology and molecular biology of phase II xenobiotic-conjugating enzymes in fish", in MALINS D.C., OSTRANDER G.K. (eds.), *Aquatic Toxicology Molecular, Biochemical, and Cellular Perspectives*, Lewis, Boca Raton, pp. 37–85, 1994.

[GIL 99] GILLEPSIE R.B., GUTTMAN S.I., "Chemical-induced changes in the genetic structure of populations: effects on allozymes", in FORBES V.E. (ed.), *Genetics and Ecotoxicology*, Taylor and Francis, New York, pp. 55–77, 1999.

[GIL 03] GILBERTSON M.K., HAFFNER G.D., DROUILLARD K.G., *et al.*, "Immunosuppression in the northern leopard frog (*Rana pipiens*) induced by pesticide exposure", *Environmental Toxicology and Chemistry*, vol. 22, no. 1, pp. 101 110, 2003.

[GOI 13] GOIN T., ARMITAGE J., COUSINS I., *et al.*, "Influence of global climate change on chemical fate and bioaccumulation: the role of multimedia models", *Environmental Toxicology and Chemistry*, vol. 32, no. 1, pp. 20–31, 2013.

[GOK 98] GØKSOYR A., HUSOY A.M., "Immunochemical approaches to studies of CYP1A localization and induction by xenobiotics in fish", in BRAUNBECK T., HINTON D.E., STREIT B. (eds.), *Fish Ecotoxicology*, Birkhäuser Verlag, Berlin, pp. 165–202, 1998.

[GOR 03] GORDON C.J., "Role of environmental stress in the physiological response to chemical toxicants", *Environmental Research*, vol. 92, no. 1, pp. 1–7, 2003.

[GOU 12] GOULLETQUER P., GROS P., WEBER J., *et al.*, *Biodiversité en environnement marin*, Quae, Versailles, 2012.

[GOU 13] GOURLAY-FRANCÉ C., TUSSEAU-VUILLEMIN M., "Bioavailability of contaminants", in FÉRARD J., BLAISE C. (eds.), *Encyclopedia of Aquatic Ecotoxicology*, Springer-Verlag, Berlin, Heidelberg, pp. 181–191, 2013.

[GRE 97] GREGG J.C., FLEEGER J.W., CARMAN K.R., "Effects of suspended, diesel-contaminated sediment on feeding rate in the darter goby, *Gobionellus boleosoma* (Teleostei: Gobiidae)", *Marine Pollution Bulletin*, vol. 34, pp. 269–275, 1997.

[GRO 06] GROFFMAN P., BARON J., BLETT T., *et al.*, "Ecological thresholds: the key to successful environmental management or an important concept with no practical application?", *Ecosystems*, vol. 9, pp. 1–13, 2006.

[GRO 99] GROENENDIJK D., KRAAK M., ADMIRAAL W., "Efficient shedding of accumulated metals during metamorphosis in metal-adapted populations of the midge Chironomus riparius", *Environmental Toxicology and Chemistry*, vol. 18, pp. 1225–1231, 1999.

[HAH 04] HAHN M.E., KARCHNER S.I., FRANKS D.G., *et al.*, "Aryl hydrocarbon receptor polymorphisms and dioxin resistance in Atlantic killifish (*Fundulus heteroclitus*)", *Pharmacogenetics*, vol. 14, pp. 131–43, 2004.

[HAH 05] HAHN M.E., FRANKS D.G., KARCHNER S.I., *et al.*, Mechanism of PCB- and dioxin-resistance in fish in the Hudson River estuary: role of receptor polymorphisms, Final Report, Hudson River Foundation Grant, New York, available at www.hudsonriver.org/ls/, p. 215, 2005.

[HAM 95] HAM L., QUINN R., PASCOE D., "Effects of cadmium on the predator–prey interaction between the turbellarian *Dendrocoelum lacteum* (Muller, 1774) and the isopod crustacean *Asellus aquaticus* (L)", *Archives of Environmental Contamination and Toxicology*, vol. 29, pp. 358–365, 1995.

[HAR 04] HARPER-ARABIE R.M., WIRTH E.F., FULTON M.H., *et al.*, "Protective effects of allozyme genotype during chemical exposure in the grass shrimp *Palaemonetes pugio*", *Aquatic Toxicology*, vol. 70, pp. 41–54, 2004.

[HAU 03] HAUSER L., ROCAP G., LARIMER F.W., *et al.*, "Molecular/cellular processes and the population genetics of a species", in LAWRENCE A.J., HEMINGWAY K.L. (eds.), *Effects of Pollution on Fish, Molecular Effects and Population Responses*, Blackwell Science, Oxford, pp. 256–288, 2003.

[HAV 93] HAVENS K.E., HANAZATO T., " ooplankton community responses to chemical stressors – a comparison of results from acidification and pesticides contamination research", *Environmental Pollution*, vol. 82, no. 3, pp. 277–288, 1993.

[HAV 95] HAVENS K.E., "Insecticide (Carbaryl, 1-Napthyl-N-Methylcarbamate) effects on a freshwater plankton community: zooplankton size, biomass, and algal abundance", *Water Air & Soil Pollution*, vol. 84, pp. 1–10, 1995.

[HEA 94] HEATH S., BENNETT W.A., KENNEDY J., *et al.*, "Heat and cold tolerance of the fathead minnow, *Pimephales promelas*, exposed to the synthetic pyrethroid cyfluthrin", *Canadian Journal of Fisheries and Aquatic Sciences*, vol. 51, no. 2, pp. 437–440, 1994.

[HEB 96] HEBERT P.D., LUIKER M.M., "Genetic effects of contaminant exposure – towards an assessment of impacts on animal populations", *Science of the Total Environment*, vol. 191, pp. 23–58, 1996.

[HEI 97] HEITHAUS M.R., LAUSHMAN R.H., "Genetic variation and conservation of stream fishes: influence of ecology, life history, and water quality", *Canadian Journal of Fisheries and Aquatic Sciences*, vol. 54, pp. 1822–1836, 1997.

[HEI 00] HEINONEN J., KUKKONEN J.V.K., HOLOPAINEN I.J., "Toxicokinetics of 2, 4, 5-trichlorophenol and benzo(a)pyrene in the clam Pisidium amnicum: effects of seasonal temperatures and trematode parasites", *Archives of Environmental Contamination and Toxicology*, vol. 39, pp. 352–359, 2000.

[HEM 07] HEMMER-HANSEN J., NIELSEN E.E., FRYDENBERG J., *et al.*, "Adaptive divergence in a high gene flow environment: Hsc70 variation in the European flounder (*Platichthys flesus* L.)", *Heredity*, vol. 99, pp. 592–600, 2007.

[HER 89] HERMENS J., "Quantitative structure-activity relationships for environmental pollutants", *Handbook of Environmental Chemistry*, Springer-Verlag, Berlin, pp. 111–162, 1989.

[HEU 01] HEUGENS E.H.W., HENDRIKS A.J., DEKKER T., *et al.*, "A review of the effects of multiple stressors on aquatic organisms and analysis of uncertainty factors for use in risk assessment", *Critical Reviews in Toxicology*, vol. 31, no. 3, pp. 247–284, 2001.

[HIN 98] HINKLE-CONN C., FLEEGER J.W., GREGG J.C., *et al.*, "The effect of sediment-bound polycyclic aromatic hydrocarbons on feeding behavior in juvenile spot (*Leiostomus xanthurus* Lace´pe`de: Pisces)", *Journal of Experimental Marine Biology and Ecology*, vol. 227, pp. 113–132, 1998.

[HOL 10] HOLMSTRUP M., BINDESBOL A.M., OOSTINGH G., *et al.*, "Interactions between effects of environmental chemicals and naturals stressors: a review", *Science of the Total Environment*, vol. 408, pp. 3746–3762, 2010.

[HOO 12] HOOPER M.J., ANKLEY G.T., CRISTOL D.A., *et al.*, "Interactions between chemical and climate stressors: a role for mechanistic toxicology in assessing climate change risks", *Environmental Toxicology and Chemistry*, vol. 32, pp. 32–48, 2012.

[IBR 98] IBRAHIM H., KHEIR R., HELMI S., *et al.*, "Effects of organophosphorus, carbamate, pyrethroid and organochlorine pesticides, and a heavy metal on survival and cholinesterase activity of Chironomus riparius Meigen", *Bulletin of Environmental Contamination and Toxicology*, vol. 60, pp. 448–455, 1998.

[IND 99] INDEHERBERG M., VAN STRAALEN N., SCHOCKAERT E., "Combinig life-history and toxicokinetic parameters to interpret differences in sensitivity to cadmium between populations of Polycelis tenuis (Plathelminthes)", *Ecotoxicology and Environmental Safety*, vol. 44, nos. 1–11, pp. 1–11, 1999.

[INT 07] INTERGOVERNMENTAL PANEL ON CLIMATE CHANGE, "Climate change 2007: the physical science basis", *Contribution of Working Group I to the Fourth Assessment Report of the Intergovernmental Panel on Climate Change*, Cambridge University Press, Cambridge, MA, 2007.

[IPC 07] INTERGOVERNMENTAL PANEL ON CLIMATE CHANGE, Climate Change: The Physical Science Basis. Contribution of Working Group I to the Fourth Assessment Report of the Intergovernmental Panel on Climate Change, Cambridge University Press, Cambridge, UK, 2007.

[JAK 96] JAK R.G., MAAS J.L., SCHOLTEN M.C.T., "Evaluation of laboratory derived toxic effect concentrations of a mixture of metals by testing fresh water plankton communities in enclosures", *Water Resource*, vol. 30, pp. 1215–1227, 1996.

[JOH 11] JOHNSTON E., "Tolerance to contaminants: evidence from chronically exposed populations of aquatic organisms", in AMIARD-TRIQUET C., RAINBOW P.S., ROMÉO M. (eds.), *Tolerance to Environmental Contaminants*, CRC Press, Boca Raton, pp. 25–48, 2011.

[KAM 96] KAMMENGA J., BUSSHERS M., VAN STRAALEN N., *et al.*, "Stress-induced fitness reduction is not determined by the most sensitive life-cycle trait", *Functional Ecology*, vol. 10, pp. 106–111, 1996.

[KAM 00] KAMMENGA J., ARTS M.J., DOROSZUK A., "Multi-generation effects at the population level: fitness maximisation and optimal allocation in a nematode", in KAMMENGA J., LASKOWSKI J. (eds.), *Demography in Ecotoxicology*, pp. 163–77, 2000.

[KAT 11] KATTWINKEL M., KÜHNE J.V., FOIT K., *et al.*, "Climate change, agricultural insecticide exposure, and risk for freshwater communities", *Ecological Applications*, vol. 21, pp. 2068–2081, 2011.

[KAU 82] KAUFMAN L.H., "Stream aufwuchs accumulation: disturbance frequency and stress resistance and resilience", *Oecologia*, vol. 52, pp. 57–63, 1982.

[KIE 02] KIESECKER J.M., "Synergism between trematode infection and pesticide exposure: a link to amphibian limb deformities in nature?", *Proceedings of the National Academy of Sciences*, vol. 99, pp. 9900-9904, 2002.

[KIF 96] KIFFNEY P.M., CLEMENTS W.H., "Size-dependent response of macro invertebrates to metals in experimental streams", *Environmental Toxicology and Chemistry*, vol. 15, pp. 1352–1356, 1996.

[KIM 10] KIM K.H., DAHMS H.U.D., RHEE J.S., *et al.*, "Expression profiles of seven glutathione S-transferase (GST) genes in cadmium-exposed river pufferfish (*Takifugu obscurus*)", *Comparative Biochemistry and Physiology, Part C*, vol. 151, pp. 99–106, 2010.

[KLA 09] KLAINE S.J., "Considerations for research on the environmental fate and effects of nanoparticles", *Environmental Toxicology and Chemistry*, vol. 28, no. 9, pp. 1787–1788, 2009.

[KLE 01] KLERKS P., MOREAU C., "Heritability of resistance to individual contaminants and to contaminant mixtures in the sheepshead minnow (*Cyprinodon variegatus*)", *Environ. Toxicol. Chem.*, vol. 20, pp. 1746–51, 2001.

[KNA 05] KNAPP C.W., CAQUET T., HANSON M.L., *et al.*, "Response of water column microbial communities to sudden exposure to deltamethrin in aquatic mesocosms", *FEMS Microbiology Ecology*, vol. 54, pp. 157–165, 2005.

[KOE 91] KOEMAN J., "From comparative physiology to toxicological risk assessment", *Comparative Biochemistry and Physiology*, vol. 100, pp. 7–10, 1991.

[KOI 92] KOIVISTO S., KETOLA M., WALLS M., "Comparison of five cladoceran species in short and long term copper exposure", *Hydrobiologia*, vol. 248, pp. 125–136, 1992.

[KOL 02] KOLPIN D.W., BARBASH J.E., GILLIOM RJ., *et al.*, Pharmaceuticals, hormones and other organic wastewater contaminants in US streams, 1999-2000: a national reconnaissance", *Environmental Science and Technology*, vol. 36, p. 1202, 2002.

[KOP 92] KOPP R.L., GUTTMAN S.I., WISSING T.E., "Genetic indicators of environmental stress in central mudminnows (*Umbra limi*) populations exposed to acid deposition in the Adirondack Mountains", *Environmental Toxicology and Chemistry*, vol. 11, pp. 665–676, 1992.

[KOP 00] KOPPENHÖFFER A.M., GREWAL P.S., KAYA H.K., "Synergism of imidacloprid and entomopathogenic nematodes against white grubs: the mechanism", *Entomologia Experimentalis et Applicata*, vol. 94, pp. 283–293, 2000.

[KRA 03] KRAEPIEL A.M.L., KELLER, K., CHIN, H.B., *et al.*, "Sources and variations of mercury in tuna", *Environmental Science & Technology*, vol. 37, no. 24, pp. 5551–5558, 2003.

[KRU 94] KRUZYNSKI G.M., BIRTWELL I.K., "A predation bioassay to quantify the ecological significance of sublethal responses of juvenile chinook salmon (*Onchorynchus tshawytscha*) to the antisapstain fungicide TCMTB", *Canadian Journal of Fisheries and Aquatic Science*, vol. 51, pp. 1780–1790, 1994.

[LAR 01] LARNO V., LAROCHE J., LAUNEY S., *et al.*, "Responses of chub (*Leuciscus cephalus*) populations to chemical stress, assessed by genetic markers, DNA damage and cytochrome P450 1A induction", *Ecotoxicology*, vol. 10, pp. 145–158, 2001.

[LAR 02] LAROCHE J., QUINIOU L., JUHEL G., *et al.*, "Genetic and physiologic responses of flounder (*Platichthys flesus*) populations to chemical contamination in estuaries", *Environmental Toxicology and Chemistry*, vol. 21, pp. 2705–2712, 2002.

[LAY 85] LAY J.P., SCHAUERTE W., PEICHL L., *et al.*, "Influence of benzene on the phytoplankton and on *Daphnia pulex* in compartments of an experimental pond", *Ecotoxicology* and *Environmental Safety*, vol. 10, pp. 218–227, 1985.

[LEB 06] LEBLANC J.C., CALIPSO study, ANSES, 2006.

[LEF 99] LEFCORT H., THOMSON S.M., COWLES E.E., *et al.*, "Ramifications of predator avoidance: predator and heavy metal-mediated competition between tadpoles and snails", *Ecological Applications*, vol. 9, pp. 1477–1489, 1999.

[LEV 09] LEVITIS D.A., LIDICKER JR. W.Z., FREUND G., "Behavioural biologists do not agree on what constitutes behaviour", *Animal Behaviour*, vol. 78, no. 1, pp. 103–110, 2009.

[LIE 11] LIESS M., BEKETOV M., "Traits and stress: keys to identify community effects of low levels of toxicants in test systems", *Ecotoxicology*, vol. 20, pp. 1328–1340, 2011.

[LIV 00] LIVINGSTONE D.R., CHIPMAN J.K., LOWE D.M., *et al.*, "Development of biomarkers to detect the effects of organic pollution on aquatic invertebrates: recent molecular, genotoxic, cellular and immunological studies on the common mussel (*Mytilus edulis* L.) and other mytilids", *International Journal of Environment and Pollution*, vol. 13, pp. 56–91, 2000.

[LOV 05] LOVEJOY T.E., HANNAH L., *Climate Change and Biodiversity*, Yale University Press, New Haven, 2005.

[LYD 99] LYDY M.J., BELDEN J.B., TERNES M.A., "Effects of temperature on the toxicity of M-parathion, chlorpyrifos, and pentachlorobenzene to Chironomus tentans", *Archives of Environmental Contamination and Toxicology*, vol. 37, no. 4, pp. 542–547, 1999.

[MA 04] MA J., HUNG H., BLANCHARD P., "How do climate fluctuations affect persistent organic pollutant distribution in North America? Evidence from a decade of air monitoring", *Environmental Science & Technology*, vol. 38, no. 9, pp. 2538–2543, 2004.

[MAC 92] MACKAY D., "Correlation of bioconcentration factors", *Environmental Science and Technology*, vol. 16, pp. 274–278, 1992.

[MAC 03] MACDONALD R.W., MACKAY D., LI Y.F., *et al.*, "How will global climate change affect risks from long-range transport of persistent organic pollutants?", Human *and Ecological* Risk Assessment, vol. 9, no. 3, pp. 643–660, 2003.

[MAC 05] MACDONALD R.W., HARNER T., FYFE J., "Recent climate change in the Arctic and its impact on contaminant pathways and interpretation of temporal trend data", *Science of the Total Environment*, vol. 342, nos. 1–3, pp. 5–86, 2005.

[MAG 97] MAGNUSON J.J., WEBSTER K.E., ASSEL R.A., *et al.*, "Potential effects of climate changes on aquatic systems: Laurentian Great Lakes and Precambrian Shield region", *Hydrological Processes*, vol. 11, pp. 825–871, 1997.

[MAR 03] MARCHAND J., TANGUY A., LAROCHE J., *et al.*, "Responses of European flounder *Platichthys flesus* populations to contamination in different estuaries along the Atlantic coast of France", *Marine Ecology Progress Series*, vol. 260, pp. 273–284, 2003.

[MAR 04] MARCHAND J., QUINIOU L., RISO R., *et al.*, "Physiological cost of tolerance to toxicants in the European flounder *Platichthys flesus*, along the French Atlantic Coast", *Aquatic Toxicology*, vol. 70, pp. 327–343, 2004.

[MAR 05] MARUYA K.A., SMALLING K.L., VETTER W., "Temperature and congener structure affect the enantioselectivity of toxaphene elimination by fish", *Environmental Science and Technology*, vol. 39, no. 11, pp. 3999–4004, 2005.

[MAR 10] MARCHAND J., EVRARD E., GUINAND B., *et al.*, "Genetic polymorphism and its potential relation to environmental stress in five populations of the European flounder *Platichthys flesus*, along the French Atlantic coast", *Marine Environmental Research*, vol. 70, pp. 201–209, 2010.

[MAY 77] MAY R.M., "Thresholds and breakpoints in ecosystems with a multiplicity of stable states", *Nature*, vol. 269, pp. 471–477, 1977.

[MEN 08] MENDOLA P., MESSER L.C., RAPPAZZO K., "Science linking environmental contaminant exposures with fertility and reproductive health impacts in the adult female", *Fertility and Sterility*, vol. 89, pp. e81–e94, 2008.

[MEY 03] MEYER J.N., WASSENBERG D.M., KARCHNER S.I., *et al.*, "Antioxidant defenses in killifish *(Fundulus heteroclitus)* exposed to contaminated sediments and model prooxidants: short term and heritable responses", *Aquatic Toxicology*, vol. 65, pp. 377–395, 2003.

[MIL 05] MILLS L.J., CHICHESTER C., "Review of evidence: are endocrine-disrupting chemicals in the aquatic environment impacting fish populations?", *Science of the Total Environment*, vol. 343, pp. 1–34, 2005.

[MOE 13] MOE S., DE SCHAMPHELAERE K., CLEMENTS W., *et al.*, "Combined and interactive effects of global climate change and toxicants on populations and communities", *Environmental Toxicology and Chemistry*, vol. 32, no. 1, pp. 49–61, 2013.

[MON 95] MONSERRAT J., BIANCHINI A., "Effects of temperature and salinity on the toxicity of a commercial formulation of methyl parathion to Chasmagnathus granulata (Decapoda, Grapsidae)", *Brazilian Journal of Medical and Biological Research*, vol. 28, no. 1, pp. 74–78, 1995.

[MON 14a] MONACO A., PROUZET P., *Ocean in the Earth System*, ISTE, London and John Wiley & Sons, New York, 2014.

[MON 14b] MONACO A., PROUZET P., *Land-Sea Interactions*, ISTE, London and John Wiley & Sons, New York, 2014.

[MON 14c] MONACO A., PROUZET P., *Ecosystem Sustainability and Global Change*, ISTE, London and John Wiley & Sons, New York, 2014.

[MOO 03] MOORE A., SCOTT A.P., LOWER N., *et al.*, "The effects of 4-nonylphenol and atrazine on Atlantic salmon (*Salmo salar* L) smolts", *Aquaculture*, vol. 222, nos. 1–4, pp. 253–263, 2003.

[MOU 11] MOUNEYRAC C., LEUNG P.T.Y., LEUNG K.M.Y., "Cost of tolerance", in AMIARD-TRIQUET C., RAINBOW P.S., ROMÉO M. (eds.), *Tolerance to Environmental Contaminants*, CRC Press, Boca Raton, pp. 265–297, 2011.

[MUL 02] MULVEY M., NEWMAN M.C., VOGELBEIN W., *et al.*, "Genetic structure of *Fundulus heteroclitus* from Polycyclic Aromatic Hydrocarbon (PAH)-contaminated and neighboring sites in the Elizabeth and York Rivers", *Aquatic Toxicology*, vol. 61, pp. 195–209, 2002.

[MUN 95] MUNKITTRICK K.R., MCCARTY L.S., "An integrated approach to aquatic ecosystem health: topdown, bottom-up or middle-out?", *Journal of Aquatic Ecosystem Stress and Recovery (Formerly Journal of Aquatic Ecosystem Health)*, vol. 4, pp. 77–90, 1995.

[MUR 94] MURDOCH M.H., HEBERT P.D.N., "Mitochondrial DNA diversity of brown bullhead from contaminated and relatively pristine sites in the great lakes", *Environmental Toxicology and Chemistry*, vol. 13, pp. 1281–1289, 1994.

[MUR 97] MURPHY D.L., GOOCH J.W., "EROD and CYP1A protein in channel catfish (*Ictalurus punctatus*) from an urban estuary relative to that in benzo[a]pyrene-exposed hatchery specimens", *Environmental Pollution*, vol. 95, no. 2, pp. 235–239, 1997.

[NAR 00] NARAHASHI T., "Neuroreceptors and ion channels as the basis for drug action: past, present, and future", *Journal of Pharmacology and Experimental Therapeutics*, vol. 294, no. 1, pp. 1–26, 2000.

[NOY 09] NOYES P.D., MCELWEE M.K., MILLER H.D., *et al.*, "The toxicology of climate change: environmental contaminants in a warming world", *Environmental International*, vol. 35, pp. 971–986, 2009.

[PAI 98] PAINE R.T., TEGNER M.J., JOHNSON E.A., "Compounded perturbations yield ecological surprises", *Ecosystems*, vol. 1, pp. 535–545, 1998.

[PAK 00] PAKERTON T., KONKEL W., "Application of quantitative structure-activity relationships for assessing the aquatic toxicology of phthalate esters", *Ecotoxicology and Environmental Safety*, vol. 45, pp. 61–78, 2000.

[PAL 08] PALUMBI S., MCLEOD K., GRUNBAUM D., "Ecosystems in action: lessons from marine ecology about recovery, resistance, and reversibility", *Bioscience*, vol. 58, no. 1, pp. 33–42, 2008.

[PAT 07a] PATERSON G., DROUILLARD K.G., HAFFNER G.D., "PCB elimination by yellow perch (Perca flavescens) during an annual temperature cycle", *Environmental Science and Technology*, vol. 41, pp. 824–829, 2007.

[PAT 07b] PATRA R.W., CHAPMAN J.C., LIM R.P., *et al.*, "The effects of three organic chemicals on the upper thermal tolerances of four freshwater fishes", *Environmental Toxicology and Chemistry*, vol. 26, no. 7, pp. 1454–1459, 2007.

[PEA 12] PÉAN S., Effcts dcs polluants organiques persistants sur le comportement des poissons, PhD Thesis, University of La Rochelle, 2012.

[PEA 13] PÉAN S., DAOUK T., VIGNET C., *et al.*, "Long-term dietary-exposure to non-coplanar PCBs induces behavioral disruptions in adult zebrafish and their offspring", *Neurotoxicology and Teratology*, vol. 39, pp. 45–56, 2013.

[PEN 01] PENUELAS J., FILELLA L., "Responses to a warming world", *Science*, vol. 294, pp. 793–795, 2001.

[PET 96] PETERSON C.H., KENNICUTT M.C., GREEN R.H., *et al.*, "Ecological consequences of environmental perturbations associated with offshore hydrocarbon production: a perspective on long-term exposures in the Gulf of Mexico", *Canadian Journal of Fisheries and Aquatic Science*, vol. 53, pp. 2637–2654, 1996.

[POD 01] PODEMSKI C.L., CULP J.M., "Toxicant interactions with food algae: a missing link between laboratory and field effects?", *Environmental Toxicology and Chemistry*, vol. 16, pp. 31–42, 2001.

[PRE 99] PRESTON B.L., CECCHINE G., SNELL T.W., "Effects of pentachlorophenol on predator avoidance behavior of the rotifer *Brachionus calyciflorus*", *Aquatic Toxicology*, vol. 44, pp. 201–212, 1999.

[REL 05] RELYEA R.A., "The impact of insecticides and herbicides on the biodiversity and productivity of aquatic communities", *Ecological Applications*, vol. 15, pp. 618–627, 2005.

[REL 06] RELYEA R.A., HOVERMAN J., "Assessing the ecology in ecotoxicology: a review and synthesis in freshwater systems", *Ecology Letters*, vol. 9, pp. 1157–1171, 2006.

[ROO 03] ROOT T.L., PRICE J.T., HALL K.R., *et al.*, "Fingerprints of global warming on wild animals and plants", *Nature*, vol. 421, no. 6918, pp. 57–60, 2003.

[ROU 87] ROUSH R.T., MCKENZIE J.A., "Ecological genetics of insecticide and acaricide resistance", *Annual Review of Entomology*, vol. 32, pp. 361–380, 1987.

[SAL 97] SALMINEN J., SETALA H., HAIMI J., "Regulation of decomposer community structure and decomposition processes in herbicide stressed humus soil", *Applied Soil Ecology*, vol. 6, pp. 265–274, 1997.

[SAL 01] SALMINEN J., ANH B.T., VAN GESTEL C.A.M., "Indirect effects of zinc on soil microbes via a keystone enchytaeid species", *Environmental Toxicology and Chemistry*, vol. 20, pp. 1167–1174, 2001.

[SCH 97] SCHINDLER D.W., CURTIS P.J., BAYLEY S.E., *et al.*, "Climate-induced changes in the dissolved organic carbon budgets of boreal lakes", *Biogeochemistry*, vol. 36, pp. 9–28, 1997.

[SCH 01] SCHEFFER M., CARPENTER S., FOLEY J.A., *et al.*, "Catastrophic shifts in ecosystems", *Nature*, vol. 413, pp. 591–596, 2001.

[SCH 05] SCHEYER A., GRAEFF C., MORVILLE S., *et al.*, "Analysis of some organochlorine pesticides in an urban atmosphere (Strasbourg, east of France)", *Chemosphere*, vol. 58, no. 11, pp. 1517–1524, 2005.

[SCH 07] SCHIEDEK D., SUNDELIN B., READMAN J.W., *et al.*, "Interactions between climate change and contaminants", *Marine Pollution Bulletin*, vol. 54, pp. 1845–1856, 2007.

[SCH 11] SCHAMPHELAERE K.A.C., GLAHOLT S., ASSELMAN J., *et al.*, "Will genetic adaptation of natural populations to chemical pollution result in lower or higher tolerance to future climate change?", *Integrated Environmental Assessment and Management*, vol. 7, pp. 141–143, 2011.

[SHI 13] SHIEL A., WEIS D., COSSA D., *et al.*, "Determining provenance of marine metal pollution in French bivalves using Cd, Zn and Pb isotopes", *Geochimica et Cosmochimica Acta*, vol. 121, no. 15, pp. 155–167, 2013.

[SLI 05] SLIJKERMAN D.M.E., MOREIRA-SANTOS M., JAK R.G., "Functional and structural impact of linuron on a freshwater community of primary producers: the use of immobilized algae", *Environmental Toxicology and Chemistry*, vol. 24, suppl. vol. I, no. 10, pp. 2477–2485, 2005.

[SMI 97] SMITH G.M., WEIS J.S., "Predator-prey relationships in mummichogs (*Fundulus heteroclitus* (L)): effects of living in a polluted environment", *Journal of Experimental Marine Biology and Ecology*, vol. 209, pp. 75–87, 1997.

[SOM 02] SOMERS C.M., YAUK C.L., WHITE P.A., *et al.*, "Air pollution induces heritable DNA mutations", *Proceedings of the National Academy of Sciences*, vol. 99, pp. 15904–15907, 2002.

[SPA 97] SPAWN R.L., HOAGLAND K.D., SIEGFRIED B.D., "Effects of alachlor on an algal community from a midwestern agricultural stream", *Environmental Toxicology and Chemistry*, vol. 16, pp. 785–793, 1997.

[SPR 05] SPROMBERG J., BIRGE J., "Modeling the effects of chronic toxicity on fish populations: the influence of life history strategies", *Environmental Toxicology and Chemistry*, vol. 24, pp. 1532–1540, 2005.

[TAS 14] TASSIN J., *La grande invasion*, Odile Jacob, Paris, 2014.

[TAY 95] TAYLOR E.J., MORRISON J.E., BLOCKWELL S.J., *et al.*, "Effects of lindane on the predator-prey interaction between *Hydra oligactis* Pallas and *Daphnia magna* Strauss", *Archives of Environmental Contamination Toxicology*, vol. 29, pp. 291–296, 1995.

[TAY 96] TAYLOR M., FEYEREISEN R., "Molecular biology and evolution of resistance to toxicants", *Molecular Biology and Evolution*, vol. 13, pp. 719–734, 1996.

[TEL 04] TELES M., GRAVATO C., PACHECO M., *et al.*, "Juvenile sea bass biotransformation, genotoxic and endocrine responses to β-naphthoflavone, 4-nonylphenol and 17 β-estradiol individual and combined exposures", *Chemosphere*, vol. 57, pp. 147–158, 2004.

[TEM 99] TEMARA A., GULEC I., HOLDWAY D.A., "Oil-induced disruption of foraging behaviour of the asteroid keystone predator, *Coscinasterias muricata* (Echinodermata)", *Marine Biology*, vol. 133, pp. 501–507, 1999.

[TRE 04] TREMOLADA P., FINIZIO A., VILLA S., *et al.*, "Quantitative inter-specific chemical activity relationships of pesticides in the aquatic environment", *Aquatic Toxicology*, vol. 67, pp. 87–103, 2004.

[TRI 97] TRIEBSKORN R., KÖHLER H.R., HONNEN W., *et al.*, "Induction of heat shock proteins, changes in liver ultrastructure, and alterations of fish behavior: are these biomarkers related and are they useful to reflect the state of pollution in the field?", *Journal of Aquatic Ecosystem Stress and Recovery* (Formerly Journal of Aquatic Ecosystem Health), vol. 6, no. 1, pp. 57–73, 1997.

[TUS 07] TUSSEAU-VUILLEMIN M.H., GOURLAY C., LORGEOUX C., *et al.*, "Dissolved and bioavailable contaminants in the Seine river basin", *Science of the Total Environment*, vol. 375, nos. 1–3, pp. 244–256, 2007.

[VAR 89] VARANASI U., *Metabolism of Polycyclic Aromatic Hydrocarbons in the Aquatic Environment*, CRC Press, Boca Raton, 1989.

[VAN 92] VAN LEEUWEN C., VAN DER ZANDT T., "Application of QSARs, extrapolation and equilibrium partitioning in aquatic effects assessment. I. Narcotic Industrial pollutants", *Environmental Toxicology and Chemistry*, vol. 11, pp. 267–282, 1992.

[VAN 95] VAN DONK E., PRINS H., VOOGD H.M., *et al.*, "Effects of nutrient loading and insecticide application on the ecology of *Elodea*-dominated freshwater microcosms I. Responses of plankton and zooplanktivorous insects", *Archiv für Hydrobiologie*, vol. 133, pp. 417–439, 1995.

[VAN 00] VAN DEN BRINK P.J., HATTINK J., BRANSEN F., *et al.*, "Impact of the fungicide carbendazim in freshwater microcosms. II. Zooplankton primary producers and final conclusions", *Aquatic Toxicology*, vol. 48, pp. 251–264, 2000.

[VAN 02] VAN DEN BRINK P.J., ROELSMA J., VAN NES E.H., *et al.*, "PERPEST model, a case-based reasoning approach to predict ecological risks of pesticides", *Environmental Toxicology and Chemistry*, vol. 21, pp. 2500–2506, 2002.

[VAN 08] VAN DEN BRINK P.J., "Ecological risk assessment: from book-keeping to chemical stress ecology", *Environmental Science and Technology*, vol. 42, pp. 8999–9004, 2008.

[VAN 12] VAN DEN BRINK P.J., TER BRAAK C.J.F., "Response to ''Traits and stress: keys to identify community effects of low levels of toxicants in test systems'' by Liess and Beketov (2011)", *Ecotoxicology*, vol. 21, pp. 297–299, 2012.

[VER 08] VERDONCK F.A.M.,VAN SPRANG P.A., VANROLLEGHEM P.A., "An intelligent data collection tool for chemical safety/risk assessment", *Chemosphere*, vol. 70, pp. 1818–1826, 2008.

[VIR 03] VIRGILIO M., BARONCINI N., TROMBINI C., *et al.*, "Relationships between sediments and tissue contamination and allozymic patterns in *Hediste diversicolor* (Polychaeta Nereididae) in the Pialassa lagoons (north Adriatic Sea)", *Oceanologica Acta*, vol. 26, pp. 85–92, 2003.

[WAL 96] WALLACE J.B., GRUBAUGH J.W., WHILES M.R., *et al.* "Biotic indices and stream ecosystem processes: results from an experimental study", *Ecological Applications*, vol. 6, pp. 140–151, 1996.

[WAL 00] WALLACE W.G., HOEXUM BROUWER T.M., BROUWER M., *et al.*, "Alterations in prey capture and induction of metallothioneins in grass shrimp fed cadmium-contaminated prey", *Environmental Toxicology and Chemistry*, vol. 19, pp. 962–971, 2000.

[WAN 96] WANIA F., MACKAY D., "Tracking the distribution of persistent organic pollutants", 1996", *Environmental Science and Technology*, vol. 30, no. 9, pp. A390–A396, 1996.

[WAN 01] WANG J., GRISLE S., SCHLENK D., "Effects of salinity on aldicarb toxicity in juvenile rainbow trout (*Oncorhynchus mykiss*) and striped bass (Morone saxatilis x chrysops)", *Toxicology Science*, vol. 64, no. 2, pp. 200–207, 2001.

[WAR 04] WARING C.P., MOORE A., "The effect of atrazine on Atlantic salmon (Salmo salar) smolts in freshwater and after sea water transfer", *Aquatic Toxicology*, vol. 66, no. 1, pp. 93–104, 2004.

[WEB 02] WEBER L.P., DIAMOND S.L., BANDIERA S.M., *et al.*, "Expression of HSP70 and CYP1A protein in ovary and liver of juvenile rainbow trout exposed to b-naphthoflavone", *Comparative Biochemistry and Physiology, Part C*, vol. 131, pp. 387–394, 2002.

[WEI 04] WEINER J., DE LORENZO M., FULTON M., "Relationship between uptake capacity and differential toxicity of the herbicide atrazine in selected microalgal species", *Aquatic Toxicology*, vol. 68, pp. 121–128, 2004.

[WEI 95] WEIS J.S., WEIS P., "Swimming performance and predator avoidance by mummichog (*Fundulus heteroclitus*) larvae after embryonic or larval exposure to methylmercury", *Canadian Journal of Fisheries and Aquatic Science*, vol. 52, pp. 2168–2173, 1995.

[WEI 01] WEIS J.S., SAMSON J., ZHOU T., *et al.*, "Prey capture ability of mummichogs (*Fundulus heteroclitus*) as a behavioural biomarker for contaminants in estuarine systems", *Canadian Journal of Fisheries and Aquatic Science*, vol. 58, pp. 1442–1452, 2001.

[WEI 02] WEIS J.S., "Tolerance to environmental contaminants in the mummichog, *Fundulus heteroclitus*", *Human and Ecological Risk Assessment*, vol. 8, pp. 933–953, 2002.

[WID 91] WIDDOWS J., DONKIN P., "Role of physiological energetics in ecotoxicology", *Comparative Biochemistry and Physiology, Comparative Pharmacology Toxic Endocrinology*, vol. 100, pp. 69–75, 1991.

[WHY 00] WHYTE J.J., JUNG R.E., SCHMITT C.J., *et al.*, "Ethoxyresorufin-O-deethylase (EROD) activity in fish as a biomarker of chemical exposure", *Critical Review in Toxicology*, vol. 30, pp. 347–570, 2000.

[WOI 98] WOIN P., "Short- and long-term effects of the pyrethroid insecticide fenvalerate on an invertebrate pond community", *Ecotoxicology and Environmental Safety*, vol. 41, pp. 137–156, 1998.

[WRO 05] WRONA F.J., PROWSE T.D., REIST J.D., *et al.*, "Freshwater ecosystems", *Arctic Climate Impact Assessment 2005*, Chapter 8, Cambridge University Press, New York, pp. 147–169, 2005. Available at: www.acia.uaf.edu/pages/scientific.html.

[XIE 04] XIE L., KLERKS P., "Fitness cost of resistance to cadmium in the least killifish (*Heterandria formosa*)", *Environmental Toxicology and Chemistry*, vol. 23, pp. 1499–1503, 2004.

[YAU 00] YAUK C.L., FOX G.A., MCCARRY B.E., *et al.*, "Induced minisatellite germline mutations in herring gulls (*Larus argenteus)* living near steel mills", *Mutation Research*, vol. 452, pp. 211–218, 2000.

[ZHO 99] ZHOU T., WEIS J.S., "Predator avoidance in mummichog larvae from a polluted habitat", *Journal of Fish Biology*, vol. 54, pp. 44–57 vol. 147, pp. 373–399, 2000, 1999.

2

Vulnerability and Resilience of Estuaries to Contamination by Antibiotics and Antibiotic-Resistant Bacteria: a Challenge for the Next Decade

2.1. Why does the ecosystem matter for human health in the emergence of antibiotic resistance

One of the major challenges for the next decades will be the assessment of how ecosystem changes could affect human health. The microbiological vulnerability and the resilience of the aquatic environment to contamination by fecal germs, which in industrialized countries is accompanied by pharmaceutical contamination, will be a major health concern (Millennium Ecosystem Assessment, http://milleniumassessment.org).

The impact of antibiotic release on the ecosystem still remains poorly understood: do they favor the selection of genes encoding antibiotic-resistance? Are these concentrations sufficient to affect microbial communities and their roles in biogeochemical cycles? The concentrations of antibiotics observed in the waters (in the order

---

Chapter written by Fabienne PETIT, Thierry BERTHE, Hélène BUDZINSKI, Roland LECLERCQ, Vincent CATTOIR, Antoine ANDREMONT, Kenny OBERLÉ, Anniet LAVERMAN, Erick DENAMUR.

of ng.L^{-1}) are lower than the minimal inhibiting concentration (MIC), thus, in theory, are not high enough to exert a selective pressure on microorganisms (mg.L^{-1}), or to affect the growth of algae (0.1–1 mg.L^{-1}), or fish such as the Japanese medaka (EC_{50} 100 mg.L^{-1} for sulfonamides) [SAN 10]. However, even subinhibiting concentrations (0.25–0.9 × CMI) can induce mutagenesis, interfer with the bacterial quorum sensing signal transmission *quorum sensing*[1], and change the gene expression [DAV 06, GOH 02, GUL 11, KOH 10, ZHA 13].

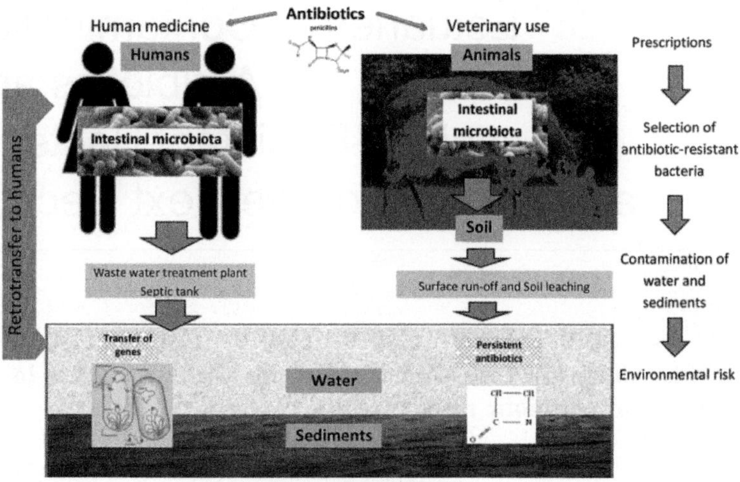

Figure 2.1. *Circulation of antibiotic-resistant microorganisms and antibiotics within the four major ecosystems: human and animals under antibiotic treatment, soil and water (see color section)*

COMMENTARY ON FIGURE 2.1.– Antibiotics prescribed in human and veterinary medicine, and fecal antibiotic-resistant bacteria, are released into the environment via urine and feces. Hence, effluent from WWTPs, and surface run-off or soil leaching, will be the major sources of contamination for waters and sediments. In an estuary environment, or in rivers, antibiotic-resistant bacteria and the more stable molecules of antibiotics are deposited on mudflats. The sediments then create an environment where (1) subinhibiting concentrations of antibiotic

1 The group of regulatory mechanisms that control the coordinated expression of certain bacterial genes within a single population of bacteria.

molecules (or their metabolites) can be reached; (2) a transfer of resistance genes from the microbial community to native species, followed by a retrotransfer to humans via pathogenic strains is possible, even if the probability is low.

Antibiotics have a unique character: with emerging contaminants of aquatic environments, their intensive use in human and animal medicine is also responsible for the emergence of antibiotic-resistant bacteria [AAR 05, SEV 02]. Since the 1950s, which saw the beginning of the large-scale use of antibiotics, pathogenic bacteria, initially sensitive to antibiotics, have very rapidly acquired resistance mechanisms, some of which were acquired via environmental bacteria. One example, the pandemic of the extended-spectrum beta-lactamases (ESBLs), whose therapeutic consequences are severe, is a consequence of a genetic transfer from environmental bacteria (e.g, CTX-M from Kluyvera spp, Poirel *et al.*, 02). Subsequently, the notion of an environmental resistome has been introduced and defined as the group of resistant genes in a given environment. Thus, in an aquatic environment, the resistome, i.e. all the resistance genes in the genome of environmental bacteria, is enriched with the genes from bacteria of human and animal origin that are discharged into this environment [WEL 13, WRI 07].

Controlling the increase of bacterial resistance to antibiotics, recognized as a major problem in public health by the World Health Organization ([OMS 01], *Millennium Ecosystem Assessment*, http://milleniumassessment. org), will be one of the leading challenges of the next decades to come.

2.2. Bacterial antibiotic resistance: a global ecological process

Bacterial resistance to antibiotics is a complex ecological phenomenon which should be investigated, by considering the circulation of micro-organisms and the corresponding resistance genes within the four major ecosystems: humans, animals, soil and water (Figure 2.1). Prescribed in human and veterinary medicine depending on the legislation in force[2], antibiotics or their metabolites are

2 The prescription antibiotics as an additive in animal food is forbidden by European Union (EU) regulations (rule no. 1831/2003).

discharged into the environment via urine and feces as a consequence of the selective pressure exerted on the intestinal microbiota of humans and animals treated with antibiotics. Consequently, in parallel with contamination by antibiotics, fecal antibiotic-resistant bacteria, notably *Escherichia coli* and *Enterococcus spp*, are released into water [GON 00, KOC 11, LAR 09, LEC 13, OBE 12, WAT 07]. Contamination of surface water is essentially due to the release of treated effluent from wastewater treatment plants (WWTP), and surface run-off or soil leaching, to which can be added uncontrolled waste from factories for pharmaceutical production and from fishing practices [JIM 12, WAT 09].

Chemical and microbiological contamination of coastal areas and estuaries reflects the anthropogenic pressure exerted on their catchments. In this aquatic environment, the fate of antibiotic-resistant bacteria will depend on their ability to overcome environmental stress (salinity, temperature and oligotrophy) or grazing by protozoa. When they are associated with organo-mineral particles, bacteria of fecal origin and the more stable molecules of antibiotics, follow particle dynamics and are deposited on mudflats, deposition zones for fine particles [COR 07, GAO 05, TAM 11, XU 09]. These mudflats, the habitat of specific microbial communities, are also zones for allochthonous bacteria [BER 08]. The antibiotic resistance genes detected in bacterial DNA extracted from highly anthropized river sediments, shows that these genes can persist in these environments [KRI 11, PEI 06]. Thus, genes prevalent in clinical strains may be disseminated to autochtonous bacteria, including phylogenetically distant bacterial species [GIR 10, GIR 11, WAL 11]. The frequency of such horizontal gene transfer is difficult to evaluate in environmental conditions. Depending on the conditions of the experiments, it ranges from 10^{-1} to 10^{-3} in optimal conditions, and from 10^{-4} to 10^{-10} in oligotrophic conditions, close to those found in the environment [BAL 88, CHA 98, MAR 06, PAU 91]. In the laboratory, a plasmid conferring co-resistance on seven antibiotics was transferred with a frequency of 2.8×10^{-3} between a strain of *E. coli* and a strain of *Aeromonas*, both isolated in the Seine Estuary [LAR 10]. These genetic transfers would therefore be possible at high cellular densities, likely to be reached in deposition zones of antibiotic-resistant bacteria, such as mudflats, periphytons, and indeed filter feeders, chronically exposed to both contamination by antibiotics and antibiotic-resistant bacteria.

Depending on human and agricultural activity in the watershed, the microbial resistome in estuarine mudflats becomes an enriched environmental resistome, where the dynamics of the horizontal gene transfer between bacteria of different phylums are still poorly understood. For the next decades, whereas the consequence of the expected increase of the demography will be accompanied by an increase in prescriptions for antibiotics, the likelihood of the retrotransfer of genes from autochtonous microorganisms to strains clinical strains, on an unknown scale of space and time, will be a major environmental concern with decisive consequences in public health [WEL 13].

2.3. Fate of contamination by antibiotics and antibiotic-resistant bacteria in estuary environments: Seine Estuary case study

2.3.1. *The Seine Estuary: one of the most anthropized estuaries in Europe*

Chemical and microbiological contamination of coastal areas and estuaries reflects the anthropogenic pressure exerted on their catchments. On the continental shelf of northwest Europe, the Seine estuary is the outlet of a drainage basin of 79,000 km^2, where 30% of the French population, 40% of the economic activity and 30% of national agricultural activity is concentrated. The microbiological quality of the Seine estuary waters is poor, and is mainly dominated, in high flow periods, by the upstream inputs of the WWTP, which treats the waste water from Paris and its suburbs (about five million inhabitants), and is located 120 km unpstream of the Dam of Poses (Figure 2.2). However, in periods of low flow, intra-estuary supplies are predominant with a contamination from tributaries close to 76% [GAR 05]. In the mouth of the estuary, where the microbiological quality of the water influences that of the mussels from the coastal zone, there is a permanent contamination by fecal bacteria, mainly due to the input of, the Risle tributary [TOU 07][3] since 1995.

2.3.2. *The fate of contamination by antibiotics, from the sources of contamination to the estuary*

The concentrations of antibiotics in water reported in the bibliography are dependent on the analytical methods and the methods of sampling. However, the maximal values (100 $\mu g\ L^{-1}$) are observed on the level of the principal sources of contamination, such as effluent from hospitals or emissions from pharmaceutical factories, whereas the values that are ten times weaker (10 $\mu g\ L^{-1}$) are detected immediately downstream from the discharged effluent of WWTPs (Table 2.1). Antibiotics, from raw effluent of the WWTPs, are degraded or eliminated in activated sludge. To that can be added photodegradation by ultraviolet rays (254 nm). In the surface waters, the lowest concentrations (10 $ng.L^{-1}$) observed can be explained by both the dilution in the receiving water and the degradation of the antibiotics or their metabolites, depending on the physico-chemical properties to each molecule ([KUM 09, TAM 08] and Table 2.1). For example, β-lactams are rarely detected since the β-lactam cycle is rapidly hydrolyzed. In the euphotic zone of the water column, photodegradation by sunlight is the major degradation pathway [LAN 09, TON 11]. However, this reaction may be partially

3 The chemical and microbiological contamination of the waters of the Seine and its estuary have been studied within the framework of two multidisciplinary research programs: the "PIREN-Seine" program (www.piren-seine.fr), which concerns the study of the hydrographic network from the headwaters of the basin to Poses (the physical limit between the river and estuarial sections), and the "Seine-Aval" (Seine-Downstream) program (http://seine-aval.crihan.fr/web/) which covers the geographic zone from Poses to the mouth of the river.

or totally inhibited depending on the concentrations of dissolved or particulate organic carbon, or of chloride, iron or bromide ions [GE 10, GIT 11, KNA 05]. On the contrary, quinolones/fluoroquinolones, sulfonamides and macrolides are stable molecules. Their properties of adsorption to particles favor their persistence in sediments, biofilms and periphytons, and in filter feeders [BRO 06, HIR 99, LIN 04, WAT 09, WUN 11].

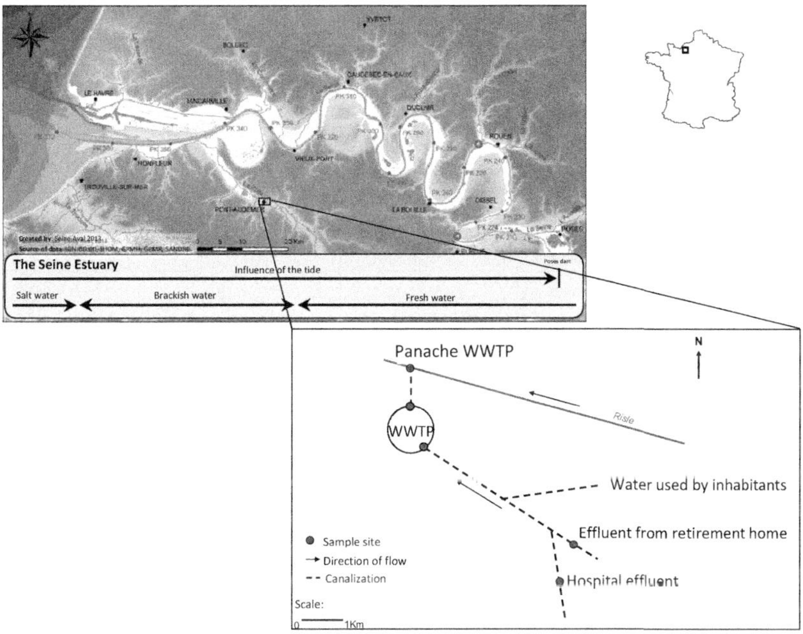

Figure 2.2. *Map of the Seine estuary: one of the most anthropized estuaries in Europe, with a population density stated in inhabitants/km² (black: 3000-6000 inh/km²; grey 500-3000 inh/km²; white < 500 inh/km²; INSEE dataset), mainly contaminated by WWTP and the upstream input from Poses; Kp, kilometric point (kp 0 at Pont Marie in Paris); and of the studied zone where the medical center is located (hospital of 87 beds and retirement home of 180 beds)-WWTP-river continuum (4km), for which the receiving river corresponds to the tributary (Risle) which flows in the mouth of the Seine estuary (see color section)*

Antibiotic	Effluent from hospitals	Untreated WWTP effluent	Treated WWTP effluent	River	Reference
– β-lactams					
- Amoxicillin	90	1,400	50	20	[WAT 09]
- Penicillin G	3,000 /	/ 10	/ 300	/ 250	[BRO 06] [WAT 09]
- Penicillin V	10	13,800	2,000	10	[WAT 09]
- Cloxacillin	/	4,600	700	/	[WAT 09]
- Cefaclor	/	6,150	1,800	200	[WAT 09]
- Cefalexin	4,100	64,000	250	100	[WAT 09]
– Quinolones and Fluoroquinolones					
- Nalidixic acid	/	200	450	50	[WAT 09]
- Ciprofloxacin	/ / 101,000 2,000 15,000 10,900	210 / / 1,000 1,100 25,800	140 / / / / 3,700	/ / / / 1,300 /	[KAR 06] [HIR 99] [LIN 04] [BRO 06] [WAT 09] [DUO 08]
- Norfloxacin	/ 90 / 15,200	/ 60 / 8,400	50 25 / 1,500	/ 30 20 /	[MIA 04] [WAT 09] [TAM 08] [DUO 08]
- Enrofloxacin	100	40	50	300	[WAT 09]
- Ofloxacin	/ 7,600 25,100	/ / 7,700	506 / 110	/ / /	[MIA 04] [LIN 04] [BRO 06]
– Tetracyclines					
- Tetracycline	/ / / 40 /	520 / / 100 /	170 nd 151 20 /	/ nd / 80 /	[KAR 06] [HIR 99] [MIA 04] [WAT 09] [BRO 06]
- Chlortetracycline	/	6	5	3	[WAT 09]
- Doxycycline	/ 6,700 200	/ / 650	46 / 150	/ / 400	[MIA 04] [LIN 04] [WAT 09]
- Oxytetracycline	/	350	70	100	[WAT 09]
– Sulfonamides					
- Sulfamethoxazole	/ / /	300 / /	200 400 243	/ 30 /	[KAR 06] [HIR 99] [MIA 04]

	12,800 1,100 100 /	/ 600 250 /	/ 310 50 /	/ 3,000 50 40	[LIN 04] [BRO 06] [WAT 09] [TAM 08]
- Sulfathiazole	nd	300	600	40	[WAT 09]
- Sulfapyridinc	/	/	34	/	[MIA 04]
– Sulfonamides					
- Sulfamethazin	/ / /	210 / /	nd nd 363	/ nd /	[KAR 06] [HIR 99] [MIA 04]
- Sulfisoxazole	/	/	34	/	[MIA 04]
- Sulfasalazine	nd	100	150	30	[WAT 09]
- Sulfadiazine	/	/	19	/	[MIA 04]
- Sulfacetamide	/	/	151	/	[MIA 04]
– Nitroimidazole					
- Metronidazole	90 200	/	/	/	[LIN 04]
- Ornidazole	/	/	/	60	[TAM 08]
– Lincosamide					
- Lincomycin	2,000 6	/ 20	/ 3	/ 1	[BRO 06] [WAT 09]
- Clindamycin	4	20	5	1	[WAT 09]
– Others					
- Trimethoprim	/ / 7,600 3,900 300 / /	330 / / 1,000 430 / /	330 660 / 1,800 10 / /	/ 200 / / 3 20 /	[KAR 06] [HIR 99] [LIN 04] [BRO 06] [WAT 09] [TAM 08]

Table 2.1. *Process of degradation and duration of the half-life (T1/2) of molecules of antibiotics in aquatic environments*

The concentrations of antibiotics observed in the Seine Estuary put in evidence this quantitative and qualitative decrease from the sources to the receiving water (Table 2.2(a)). Thus, the treated effluent from the WTTP, and the tributary impacted by effluents from a hospital (Robec), are the two principal sources of intra-estuary contamination by antibiotic molecules, with nearly 20 different antibiotic molecules, and maximal concentrations of 482 $ng.L^{-1}$ observed for clarithromycin (from the macrolide family) (Table 2.2(b)). The contamination from antibiotics from WWTP emissions is rapidly diluted in the waters of the estuary, by a factor of at least 10 depending on the molecules, as shown in concentrations measured in the urban area around Rouen, especially on the site closest to the WWTP discharge. With the exception of sulfamethoxazole (sulfonamide family), this decrease is true for all the antibiotics along the estuary especially in Caudebec (limit of the salinity), and then in the mouth of the estuary. The antibiotics predominantly found are those that are most stable in the aquatic environment: fluoroquinolones, sulfonamides and macrolides.

However, the multiplicity of intra-estuary supplies, combined with the hydrosedimentary dynamic, does not allow a clear understanding of the fate of antibiotics in this aquatic environment. A finer scale analysis of the relationship between antibiotic use and the contamination of water by antibiotics has been therefore carried out along a medical center (hospital of 87 beds and retirement home of 180 beds) – the WWTP-river continuum (4 km). The receiving river, corresponding to the last tributary (Risle), flows in the mouth of the Seine estuary (the Risle, Figure 2.2).

Contamination in water by antibiotic molecules results directly from medical prescriptions (type of molecule and epidemic period), and from the persistence in water of antibiotics and/or their metabolites (Table 2.3).

Even though penicillin and amoxicillin are the most prescribed antibiotics in hospitals and in the town's pharmacies, the main antibiotics found in the effluent from the hospitals are the molecules that are most persistent in water: the quinolones/fluoroquinolones (ofloxacin: 68 $\mu g\ L^{-1}$; pipemidic acid: 59 $\mu g\ L^{-1}$) and the macrolides, to which can be added cephalosporins (160 $\mu g\ L^{-1}$) (Table 2.3(a) and (b)). All the antibiotics prescribed at the hospital and in the town's pharmacies are found in the raw effluent of the WWTP, but the concentrations observed are low (in the order of $ng.L^{-1}$) and decrease further in the receiving water, where only the quinolones, the sulfonamides and the macrolides are detected at concentrations in the order of $ng.L^{-1}$. These results highlight that antibiotic contamination of waters correlates to the medical prescriptions (type of molecule and epidemic period), but also to the stability of antibiotics and/or their metabolites once released in the water.

Antibiotic (ng/L)	The Robec	WWTP Elbeuf	WWTP Emeraud
β-lactams			
Amoxicillin	235	-	41
Ampicillin	-	-	-
Penicillin G	-	-	-
Penicillin V	-	-	-
Oxacillin	12	-	-
Cloxacillin	40	-	-
Dicloxacillin	-	-	-
Cephalosporins			
Cefalexin	-	-	-
Cefotaxim	368	31	12
Cefpodoxine	-	-	-
Ceftiofur	-	-	-
Cefuroxime	-	-	-
Quinolones/ Fluoroquinolones			
Ciprofloxacin	114	130	265
Enrofloxacin	-	-	-
Marbofloxacin	-	-	-
Norfloxacin	-	22	165
Oxfloxacin	62	77	86
acid			
Pipemidic acid	-	-	148
Oxolinic acid	-	-	-
Flumequine	-	161	138
Tetracyclins			
Tetracycline	-	-	9
Oxytetracycline	-	-	-
Chlortetracycline	-	-	-
Doxycycline	-	-	17
Sulfonamides			
Sulfadiazine	-	342	2
Sulfadimethoxine	-	129	0
Sulfamerazine	-	-	-
Sulfamethazine	-	105	-
Sulfamethizole	-	10	4
Sulfamethoxazole	5	276	200
Sulfanilamide	-	137	67
Sulfapyridine	-	235	396
Sulfathiazole	-	-	-
Macrolides			
Azithromycin	-	208	248
Clarithromycin	3	89	482
Erythromycin	3	126	295
Josamycin	-	11	21
Roxithromycin		73	239
Spiramycin	-	39	50

-: antibiotic not detected

Table 2.2(a). *The major sources of contamination of the Seine estuary by antibiotics are: the tributary (Robec), contaminated by effluent of an hospital (1250-beds capacity) which flows at Kp 242 (urban zone of Rouen), and the two most important WWTP located in the urban zone, Rouen (Kp 247, treatment capacity 550,000 inhabitants) and Elbeuf (Kp 220 treatment capacity 100,000 inhabitants) [GAR 05, TOU 07]. A chemical analysis methodology, based on solid phase extraction coupled with liquid chromatography tandem mass spectrometry, was developped in order to detect 34 molecules of different antibiotics [OBE 12]*

Antibiotic (ng/L)	Poses	Rouen	Croisset	Docks	Bouille	Caudebec	Tancerville	Berville	Honfleur
β-lactams									
Amoxicillin	-	-	-	-	-	-	-	-	-
Ampicillin	-	-	-	-	-	-	-	-	-
Penicillin G	-	-	-	-	-	-	-	-	-
Penicillin V	-	-	-	-	-	-	-	-	-
Oxacillin	-	-	-	-	-	-	-	-	-
Cloxacillin	-	-	-	-	-	-	-	-	-
Dicloxacillin	-	-	-	-	-	-	-	-	-
Cephalosporins									
Cefalexin	-	-	-	-	-	-	-	-	-
Cefotaxime	-	-	-	-	-	-	-	-	-
Cefpodoxime	-	-	-	-	-	-	-	-	-
Cefiotur	-	-	-	-	-	-	-	-	-
Cefuroxime	-	-	-	-	-	-	-	-	-
Quinolones/ Fluoroquinolones									
Ciprofloxacin	-	-	-	-	-	-	-	-	-
Enrofloxacin	-	-	-	-	-	-	-	-	-
Marbofloxacin	-	-	-	-	-	-	-	-	-
Norfloxacin	-	-	-	-	-	-	-	-	-
Ofloxacin	3	2	2	2	1	-	1	1	1
Pipemidic acid	-	-	-	-	-	-	-	-	-
Oxilinic acid	-	-	-	-	-	-	-	-	-
Flumequine	9	9	9	8	13	7	3	6	-
Tetracyclines									
Tetracycline	-	-	-	-	-	-	-	-	-
Oxytetracycline	-	-	-	-	-	-	-	-	-
Chlortetracycline	-	-	-	-	-	-	-	-	-
Doxycycline	-	-	-	-	-	-	-	-	-
Sulfonamides									
Sulfadiazine	-	-	-	-	-	-	-	3	1
Sulfadimethoxine	1	0	1	0	0	-	0	-	-
Sulfamerazine	-	-	-	-	-	-	-	-	-
Sulfamethazine	-	-	-	-	-	-	-	-	-
Sulfamethizole	-	-	-	-	-	-	-	-	-
Sulfamethoxazole	24	5	24	16	21	51	22	56	12
Sulfanilamide	-	-	-	-	-	-	-	-	-
Sulfapyridine	12		6	5	5	16	7	14	2
Sulfathiazole	-	-	-	-	-	-	-	-	-
Macrolides									
Azithromycin	-	-	-	-	-	-	-	-	-
Clarithromycin	8	9	11	7	11	1	0	1	2
Erythromycin	20	19	24	18	20	9	2	4	3
Josamycin	-	-	-	-	-	-	-	-	-
Roxithromycin	2	3	4	2	3	1	-	0	-
Spiramycin	5	2	3	2	2	1	0	1	1

-: antibiotic not detected

Table 2.2(b). *The major sources of contamination of the Seine estuary by antibiotics are: the tributary (Robec), contaminated by effluent of an hospital (1250-beds capacity) which flows at Kp 242 (urban zone of Rouen), and the two most important WWTP located in the urban zone, Rouen (Kp 247, treatment capacity 550,000 inhabitants) and Elbeuf (Kp 220 treatment capacity 100,000 inhabitants) [GAR 05, TOU 07]. A chemical analysis methodology, based on solid phase extraction coupled with liquid chromatography tandem mass spectrometry, was developped in order to detect 34 molecules of different antibiotics [OBE 12]*

Antibiotic	Medical centre – Hospital: Concentration ($\mu g.L^{-1}$)	Hospital: % Resistant bacteria		Medical centre – Retirement home: Concentration ($\mu g.L^{-1}$)	Retirement home: % Resistant bacteria		Raw effluent WWTP: Concentration ($\mu g.L^{-1}$)	Raw effluent WWTP: % Resistant bacteria		Treated effluent from WWTP: Concentration ($\mu g.L^{-1}$)	Treated effluent from WWTP: % Resistant bacteria		River (Risle): Concentration ($\mu g.L^{-1}$)	River (Risle): % Resistant bacteria	
Ampicillin	0.0	10	100[b]	0.7	3[illegible][a]	87.5[b]	0.0	6.0[a]	19.1[a]	0.0	10.0[a]	19.0[b]	0.0	16[a]	4.0[b]
Other β-lactams	161	-	-	0.0	-	-	0.0	-	-	0.0	-	-	0.0	-	-
Ciprofloxacin	73	8	75[b]	0.0	42[a]	100.0[b]	0.0	16[a]	15.6[b]	0.04	4.0[a]	17.0[b]	0.0	4.0[a]	5.0[b]
Other Quinolones	68	-	-	59	-	-	0.4	-	-	0.1	-	-	0.001	-	-
Tetracycline	0.0	16[a]	0.0[b]	0.0	33[a]	21.0[b]	0.08	34[a]	28.0[b]	0.01	20[a]	26.0[b]	0.0	18[a]	24.0[b]
Other Tetracycines	1.0	-	-	0.0	-	-	0.07	-	-	0.0	-	-	0.0	-	-
Erythromycin	0.4	-	72[b]	0.0	-	87.0[b]	0.4	-	77.0[b]	0.05	-	80.0[b]	0.002	-	72.0[b]
Other macrolides	7.5	-	-	21.9	-	-	3.1	-	-	1.6	-	-	0.005	-	-
Cephalosporin	160	2[a]	-	0.0	6[a]	-	0.62	6.0[a]	-	0.04	2.0[a]	-	0.000	2.0[a]	-
Sulfonamides	0.0	2[a]	-	0.006	24[a]	-	0.924	16[a]	-	0.715	8.0[a]	-	0.005	6.0[a]	-

a: *E. coli*; b: *Enterococcus*; -: resistance untested.

Table 2.3(a). *Water contamination by antibiotics and fecal antibiotic-resistant bacteria, Escherichia coli and Enterococcus, along the continuum medical center – WWTP (collecting the waste waters from the medical center and from a community of 9,058 inhabitants) – River (Risle), during the winter period and an high epidemic period (December 2009). The medical center is composed of a hospital (87 beds, residence time from 4 to 28 days) and a retirement home (180 beds, resisdence time an average of ten years), [LEC 13, OBE 12]*

Whatever the antibiotic, the concentrations observed along the *continuum* are lowest than the MICs responsible for the selection of antibiotic-resistant bacteria, which are in the order of 0.032 μg mL^{-1} to 256 μg mL^{-1}, and remain inferior to the subinhibiting concentrations (0.9x CMI to 0.25x CMI), for which a possible effect on bacterial physiology has been described [DAV 06, GUL 11, KOH 10, ZHA 13].

Antibiotic (g)	Hospital	Retirement home	Commune
– Glycopeptide			
- Vancomycin	0	0	n.a
– β-lactams			
- Amoxicillin	1,930	544	258,430
- Cloxacillin	0	16	736
- Ampicillin	0	0	0
– Tetracyclines			
- Tetracycline	0	0	0
- Doxycycline	1.1	0	1,562
– Fluoroquinolones			
- Ciprofloxacin	115	0	980
– Macrolides			
- Erythromycin	75	20	n.a

Table 2.3(b). *Antibiotic consumption a antibiotic consumption 1month before sampling, b antibiotic consumption 6 months before sampling based on the sales of antibiotics for all the pharmacies in the studied zone, NA, not available (from [LEC 13])*

2.3.3. *Occurence of antibiotic-resistant* Escherichia coli *and* Enterococcus *in the Seine estuary*

In parallel with the contamination by antibiotics, a permanent high level of antibiotic-resistant *E. coli* has been observed in the Seine estuary ([LAR 09]; Table 2.4). A high proportion of these strains are multiresistant to antibiotics (60.5% to 80%), part of which (11.1%) harbor an integron (Table 2.5). The class 1 integrons are found among the *E. coli* strains all along the estuary, whereas only the strains isolated in the urban zone (Rouen) harbor class 2 integrons. The class

1[4] integrons are found among the *E. coli* strains all along the estuary, whereas only the strains isolated in the urban zone around Rouen possess class 2 integrons. These supplies of antibiotic-resistant *E. coli* harboring class 1 integrons are mainly of human origin via the treated effluent from the WWTP [LAR 09].

	Station	pk (km)	*Enterococcus* spp[b]		*Escherichia coli*[c]	
			min – max [a]	*Geom. mean. (%)*[e]	*min – max*[a]	*Geom.mean. (%)*[e]
Upstream input	*Poses*	202	< 1– 3,000	217 (54.2)	34 – 54,130	835 (54.5)
Urban zone	*Rouen*	243	36–9,300	570 (57.1)	292 – 23,690	2,194 (95.7)
	Croisset	246.6	57 – 23 000	635 (59.1)	350 – 26,700	2,247 (96.3)
	Docks	251.3	36 – 9,300	522 (77.3)	160 – 16,620	2,183 (95.5)
	Bouille	259.7	15 – 4,300	266 (65.2)	293 – 6,520	1,195 (82.1)
Salinity boundary	*Caudebec*	310.5	7 – 2,300	111 (36.4)	38 – 1,126	181 (22.7)
Mouth of the estuary	*Tancarville*	337	10 – 2,300	110 (33.3)	36 – 2,472	154 (15.2)
	Berville	346	23 – 600	86 (na)	na	na
	Honfleur	355.8	3 – 11,454	171 (54.2)	36 – 4,300	338 (28.1)

Table 2.4(a). *Microbiological quality of water along the Seine estuary: a) abundance of bacteria (UFC.100 ml^{-1}) in E. coli (2000–2005, thirty sampling campaigns); na: not analyzed; percentage of campaigns for which the abundance of E. coli and Enterococci is greater than the values given in the European directive (2006/7/EC): respectively 185 and 500 UFC.100 mL^{-1} [TOU 07]*

Water contamination by antibiotic-resistant bacteria was also; investigated along the medical center-WWTP-river continuum. The results show that the percentages of strains of *E. coli* resistant to at

4 Integrons, are genetic supports responsible for the capture of antibiotic resistance genes. They play a major role in the dissemination of antibiotic resistance genes, mainly among Gram-negative bacteria.

least one antibiotic, and multiple-resistant strains (from three to eight strains), are greater in the effluent of the retirement home than in the effluent of the hospital (Tables 2.3).

Stations	pK (km)	Number of strains of *E. coli* tested	% of antibiotic-resistant isolates[a]	
			min–max	*Geom.mean*
Poses	202	113	16.0–50.0	30.2
Rouen	243	77	36.8-39.6	38.2
Le Croisset	246.6	104	24.5-94.7	56.6
La Bouille	260	92	31.0-68.7	49.6
Caudebec	310	94	13.3-58.3	35.7
Tancarville-Honfleur	337 – 356	174	44.2-60.0	50.1
WWTP Emeraude	247	123	46.1-50.0	48.3

Table 2.4(b). *Proportion of E. coli isolates (%) resistant to at least one antibiotic in water, along the Seine estuary (January 2006, N = 279) [LAR 09]*

Stations	Kp (km)	No. of *E. coli* isolated	% resistant isolates [a]	% of isolates possessing an integron				Total
				intI 1	intI 2	intI 1 + 2	intI 3	
Poses	202	16	25.0	6.2	0	0	0	6.2
Rouen	243	58	39.6	8.6	5.2	0	0	13.8
Le Croisset	246.6	22	54.5	9.1	4.5	0	0	13.6
La Bouille	260	35	42.8	2.7	0	0	0	2.7
Caudebec	310	53	45.3	3.8	0	0	0	5.7
Tancarville-Honfleur	337–356	95	44.2	14.7	0	1.1	0	15.8
Total		279	43.0	8.9	1.4	0.7	0	11.0

Table 2.4(c). *Occurrence of E. coli strains carrying an integron (intI1, intI2 or intI3) in water, along the Seine Estuary (January 2006, N = 279) [LAR 09]*

However, no simple relationship exists between the antibiotic-resistance phenotypes of *E. coli* and *Enterococcus*, isolated in the waters, with contamination of the environment by antibiotics and prescriptions. The greatest antibiotic-resistance observed in *E. coli* strains, isolated from the effluent of the retirement home, corresponds to the most prescribed antibiotics: quinolones/ fluoroquinolones (42%) and penicillins (38%). A significant proportion of the *E. coli* strains are resistant to sulfonamides (24%), and tetracyclines (24%), even though these antibiotics are not prescribed (Table 2.3(a)). The presence of cassettes contained in the class 1 integrons are detected in the genomes of 12/15 of the strains. If, for each antibiotic prescribed, there is a corresponding resistance phenotype among the *E. coli* strains released into the water, the phenotypes for multiresistance observed are explained by the presence of integrons [CAM 10, STA 12].

The epidemiological strains from the hospital, carrying specific genetic supports (integrons, a gene for resistance to *erm* macrolides) decreased along the continuum in favor of strains better adapted to the environment (Figure 2.3). Thus, *E. coli* resistant to at least one antibiotic and those carrying the class 1 integron, decreased significantly along the continuum (p-value < 0.001) from the retirement home effluents to the river (Table 2.2, Figure 2.3, [OBE 12].

Among the *Enterococci* population, but mainly identified as *Enterococcus faecium*, all isolates from the hospital were multi-resistant, contained *erm*(B) and *mef*(A) and belonged to hospital-adapted clonal complex 17(CC17) [LEC 13]. It decreases in the treated effluent from the waste water treatment plant (19% and 17%) to reach 4% and 5% in the waters of the Risle. Interestingly, while the proportion of *Enterococcus faecium* resistant to erythromycin is greater than 70% along the the continuum, only the hospital isolates exhibit a high level of resistance. The prevalence of the *erm*(B) gene reaches 75% of the hospital isolates and only 6.7% of the strains isolated in the river (Figure 2.4). Similarly, the occurrence of the clonal complex "CC17", sequence of DNA present among the epidemic hospital strains (Figure 5; [LEA 06]), is maximal among the strains of *Enterococcus faecium* isolated in the hospital effluents, and then decrease along the continuum. This study shows a preferential disappearance of epidemiologic strains of *E. coli* and *Enterococcus faecium* in favor of other sub-populations of these bacteria species less resistant to antibiotic and probably better adapted to the aquatic environment [BER 13, PHA 13, RAT 10].

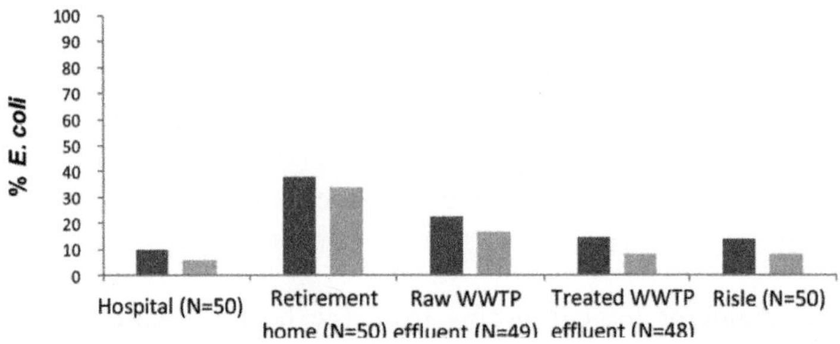

Figure 2.3(a). *Occurence of multiple-resistant strains of E. coli along the medical center-WWT river continuum: percentage of isolates of E. coli resistant to at least three antibiotics (dark gray), percentage of isolates of E. coli carrying class1 integron (light gray) medical center – WWTP river [ORB 11] (see color section)*

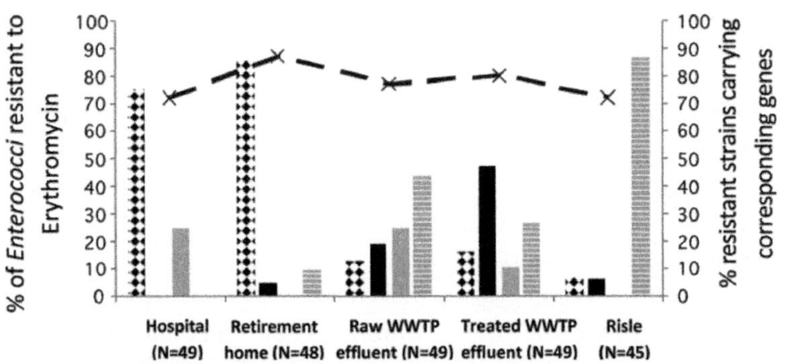

Figure 2.3(b). *Occurence of Enterococcus strains resistant to erythromycin, along medical center – WWTP – m continuum: (- - -), percentage of Enterococcus strains resistant to erythromycin (macrolides family), percentage of strains of Enterococcus resistant to erythromycin carrying corresponding genes ermB* ⁙; *mefA n*■, *ermB+mefA n*■, *unknown genes*≡ *(Leclercq et al., 2013) (see color section)*

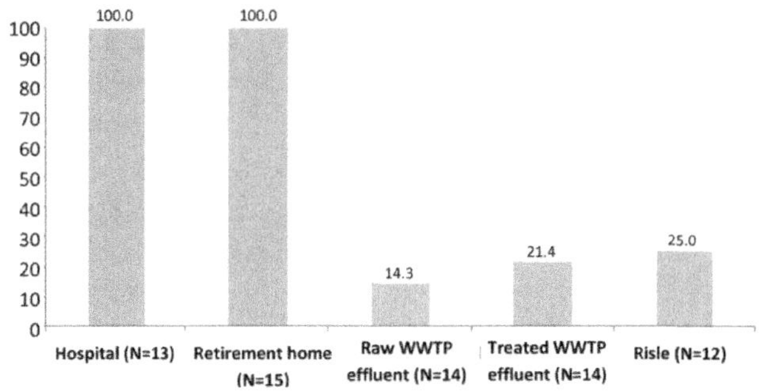

Figure 2.3(c). *Occurrence of epidemiologic strains of Enterococcus, carrying the DNA sequence marker CC17, within the population of Enterococcus faecium, along the medical center-WWTP-river continuum (Leclercq et al., 2013) [LEC 13]*

2.4. Estuary sediments: a vulnerable environment?

2.4.1. *The resistome in estuary sediments*

Bacteria of fecal origin and the chemical contaminants associated with them (trace metals, organic contaminants and antibiotics) are deposited on the mudflats, where bacteria rapidly lose their cultivability. Thus, these mudflats could be a suitable area for remedial capacity [BER 08]. However, the molecular detection of class 1 integrons in the mudflats along the Risle-estuary transect, where no culturable bacteria is present, suggests that the sediments of an anthropized estuary could be a favorable environment for the persistence of the DNA of fecal bacteria, and therefore, potentially for antibiotic resistance genes. This area could therefore constitute a favorable environment for the horizontal gene transfer mainly by transformation and/or transposition[5].

2.4.2. *Impact of contamination by antibiotics on the functional microbial community of sediments*

Whereas concentrations of antibiotics are *a priori* too low to exert a selective pressure on the environmental biota or the microbiota in

5 Incorporation of nucleic acids synthesized outside the genome.

the anthropized river sediments, the most persistent antibiotics are detectable (macrolides, quinolones, sulfonamides and tetracyclines) at concentrations that can reach 100 $\mu g.kg^{-1}$ for antibiotics from the quinolones family [MAS 10]. In the sediments of the Seine, antibiotics can persist for more than 30 years, at concentrations that can reach 32 $\mu g.kg^{-1}$ for quinolones, 15 $\mu g.kg^{-1}$ for sulfonamides and 20 $\mu g.kg^{-1}$ for nalidixic acid [TAM 11]. Laboratory experiments have shown that a continuous exposure of sediments to concentrations of antibiotics of 1 $mg.L^{-1}$ and 250 $mg.L^{-1}$ (erythromycin, clarithromycin, amoxicillin and oxytetracycline) inhibits the activity of nitrifying and denitrifying microbial communities [LUI 01]. At the lowest concentrations of vancomycin or flumequine (100 and 1,000 $ng.L^{-1}$), the diversity of microbial communities is modified, whereas the rate of nitrate reduction in denitrifying bacterial communities does not vary significantly [YAN 13]. This phenomenon of resilience, which is explained by a functional redundancy within microbial community, has also been observed among sulfate-reducing microbial communities present in the sediments contaminated by antibiotics in marine fish farms [HAN 92].

2.5. Vulnerability and resilience in the estuary environment

In the decades to come, estuaries will be exposed to a great chemical and microbial contamination, in correlation to the increase of the demography and human activity on their watershed. The study carried out in the Seine Estuary, and at the small-scale of a hospital-WWTP-river continuum, shows the vulnerability, but also the resilience, of this environment, to the contamination by antibiotics and antibiotic-resistant bacteria.

Thus, both the concentrations of antibiotic and the occurence of antibiotic-resistant bacteria greatly decrease during their transfer from the main source of contamination (WWTP, hospital) to the estuary. The substantial decrease of antibiotic concentration along this continuum is mainly due to degradation or elimination by WWTP treatments, and the subsequent dilution in the receiving environment. Whatever the antibiotic, the concentrations observed are lower than the minimal inhibitory concentrations (MIC ranging from 0.032 μg mL-1 to 256 μg mL-1) responsible for the selection of antibiotic-resistant bacteria, demonstrating the resilience of this water environment to contamination by antibiotics [OBE 12, LEC 13]. However, possible effects on bacterial physiology have been described for subinhibitory or sublethal concentrations (0.9x CMI to 0.25x CMI) (Davies *et al.*, 2006, Kohanski *et al.*, 2010). Indeed, the most stable molecules, such as the quinolones, the macrolides and the sulfonamides, which persist longer in water, could accumulate in environments such as the biofilms of periphytons or estuary mudflats.

The occurrence of antibiotic-resistant *E. coli* and *Enterococcus* strains released in waters results from the selective pressure exerted on the intestinal microbiota of humans during antibiotic treatment. In the water, no relationship between the phenotypes for resistance in fecal bacteria and contamination by antibiotics exists, since only the most stable molecules are detectable and the bacteria can harbor integrons that confer resistance to several antibiotics. However, the abundance of antibiotic-resistant fecal bacteria decreases during their transfer from the source to the estuary, mainly due to a more important decay of hospital strains, that harbor gene implicated in spread of antibiotic resistance (integrons, *erm* gene), in favor of strains less resistant to antibiotic and probably better adapted to the environment. These results also underline the resilience capacity of this aquatic environment [BER 13, LEC 13].

However, the estuarine sediments chronically exposed to multiple chemical contaminants, including antibiotics, to which are added supplies of antibiotic-resistant bacteria, are vulnerable environments. If indeed in the sediments exposed to contamination by antibiotics, the degradation of the organic matter by the denitrifying or sulfate-reducing communities seems barely affected due to the functional redundancy of the microbial communities, this effect of exposure to multiple contamination is not described in the literature. In estuarine mudflats, the enrichment of the bacterial resistome of the sediments with antibiotic-resistant genes from fecal bacteria, in an environment where subinhibiting concentrations of antibiotics could be reached, thus making this area as a putative hot spot zone favorable to the transfer of antibiotic-resistance gene within the microbial communities. These mudflats could be especially vulnerable when they are located close to the discharged of the major source of will be even more vulnerable when they are closer to the sources of contamination (WWTP, hospital emissions, industrial emissions or fishing).

2.6. Acknowledgments

These studies was supported by the project FLASH (GIP Seine-Aval/ EC2CO CNRS), SFR SCALE and by research grants from the Haute Normandie regional council (France) for the PhD of Kenny Oberlé, Mehdy Ratajczak and Emilie Laroche. We thank Michel Simon, Caroline Bance and Michel Auzou for excellent technical assistance. We thank also Michel Leroux, Aurélie Lamy, Yvon Goarvot, Sophie Coté, and downtown pharmacists for antibiotic consumption data, and the people that let us access to the medical center and to the WWTP.

2.7. Bibliography

[AAR 05] AARESTRUP F.M., "Veterinary drug usage and antimicrobial resistance in bacteria of animal origin", *Basic Clinical Pharmacology Toxicology*, vol. 96, pp. 271–281, 2005.

[BAL 88] BALE M.J., DAY M.J., FRY J.C., "Novel method for studying plasmid transfer in undisturbed river epilithon", *Applied and Environmental Microbiology*, vol. 54, no. 11, pp. 2756–2758, 1988.

[BER 08] BERTHE T., TOURON A., LELOUP J., *et al.*, "Faecal-indicator bacteria and sedimentary processes in estuarine mudflats (Seine, France)", *Marine Pollution Bulletin*, vol. 57, nos. 1–5, pp. 59–67, 2008.

[BER 13] BERTHE T., RATAJCZAK M., CLERMONT O., *et al.*, "Evidence for co-existence of distinct Escherichia coli population regarding their survival in aquatic environment", *Applied Environmental Microbiology*, vol. 79, pp. 4684–4693, 2013.

[BRO 06] BROWN K.D., KULIS J., THOMSON B., *et al.*, "Occurrence of antibiotics in hospital, residential, and dairy effluent, municipal wastewater, and the Rio Grande in New Mexico", *Science of the Total Environment*, vol. 366, nos. 2–3, pp. 772–783, 2006.

[CAM 10] CAMBRAY G., GUEROUT A.M., MAZEL D., "Integrons", *Annual Review of Genetics*, vol. 44, pp. 141–166, 2010.

[CHA 98] CHANDRASEKARAN S., VENKATESH B., LALITHAKUMARI D., "Transfer and expression of a multiple antibiotic resistance plasmid in marine bacteria", *Current Microbiology*, vol. 37, no. 5, pp. 347–351, 1998.

[COR 07] CÓRDOVA-KREYLOS A.L., SCOW K.M., "Effects of ciprofloxacin on salt marsh sediment microbial communities", *The ISME Journal*, vol. 1, no. 7, pp. 585–595, 2007.

[DAV 06] DAVIES J., SPIEGELMAN G.B., YIM G., "The world of subinhibitory antibiotic concentrations", *Current Opinion in Microbiology*, vol. 9, no. 5, pp. 445–453, 2006.

[DUO 08] DUONG H.A., PHAM N.H., NGUYEN H.T., *et al.*, "Occurrence, fate and antibiotic resistance of fluoroquinolone antibacterials in hospital wastewaters in Hanoi, Vietnam", *Chemosphere*, vol. 72, no. 6, pp. 968–973, 2008.

[GAO 05] GAO J., PEDERSEN J.A., "Adsorption of sulfonamide antimicrobial agents to clay minerals", *Environmental Science & Technology*, vol. 39, no. 24, pp. 9509–9516, 2005.

[GAR 05] GARCIA-ARMISEN T., TOURON A., PETIT F., *et al.*, "Sources of faecal contamination in the Seine estuary (France)", *Estuaries and Coasts*, vol. 28, no. 4, pp. 627–633, 2005.

[GE 10] GE L., CHEN J., WEI X., *et al.*, "Aquatic photochemistry of fluoroquinolone antibiotics: kinetics, pathways, and multivariate effects of main water constituents", *Environmental Science & Technology*, vol. 44, no. 7, pp. 2400–2405, 2010.

[GIT 11] GITHINJI L.J.M., MUSEY M.K., ANKUMAH R.O., "Evaluation of the fate of ciprofloxacin and amoxicillin in domestic wastewater", *Water, Air & Soil Pollution*, vol. 219, no. 1, pp. 191–201, 2011.

[GIR 10] GIRLICH D., POIREL L., NORDMANN P., "Novel ambler class A carbapenem-hydrolyzing beta-lactamase from a *Pseudomonas fluorescens* isolate from the Seine River, Paris, France", *Antimicrobial Agents Chemotherapy*, vol. 54, pp. 328–332, 2010.

[GIR 11] GIRLICH D., POIREL L., NORDMANN P., "Diversity of clavulanic acid-inhibited extended-spectrum beta-lactamases in *Aeromonas* spp. from the Seine River, Paris, France", *Antimicrobial Agents Chemotheraterapy*, vol. 55, pp. 1256–1261, 2011.

[GOH 02] GOH E.B., YIM G., TSUI W., *et al.*, "Transcriptional modulation of bacterial gene expression by subinhibitory concentrations of antibiotics", *Proceedings of the National Academy of Sciences of the United States of America*, vol. 99, no. 26, p. 17025, 2002.

[GON 00] GONI-URRIZA M., CAPDEPUY M., ARPIN C., *et al.*, "Impact of an urban effluent on antibiotic resistance of riverine Enterobacteriaceae and *Aeromonas* spp", *Applied Environmental Microbiology*, vol. 66, pp. 125–132, 2000.

[GUL 11] GULLBERG E., CAO S., BERG O.G., *et al.*, "Selection of resistant bacteria at very low antibiotic concentrations", *PLoS Pathogens*, vol. 7, no. 7, p. e1002158, 2011.

[HAN 92] HANSEN P.K., LUNESTAD B.T., SAMUELSEN O.B., "Effects of oxytetracycline, oxolinic acid, and flumequine on bacteria in an artificial marine fish farm sediment", *Canadian Journal of Microbiology*, vol. 38, no. 12, pp. 1307–1312, 1992.

[HIR 99] HIRSCH R., TERNES T., HABERER K., *et al.*, "Occurrence of antibiotics in the aquatic environment", *The Science of the Total Environment*, vol. 225, no. 1–2, pp. 109–118, 1999.

[JIM 12] JIMÉNEZ C., NARVAEZ J.F., PENUELA G.A., "Pharmaceutical products in the environment: sources, effects and risks", *Vitae*, vol. 19, no. 1, pp. 92–108, 2012.

[KAR 06] KARTHIKEYAN K.G., MEYER M.T., "Occurrence of antibiotics in wastewater treatment facilities in Wisconsin, USA", *Science of the Total Environment*, vol. 361, nos. 1–3, pp. 196–207, 2006.

[KNA 05] KNAPP C.W., CARDOZA L.A., HAWES J.N., *et al.*, "Fate and effects of enrofloxacin in aquatic systems under different light conditions", *Environmental Science & Technology*, vol. 39, no. 23, pp. 9140–9146, 2005.

[KOC 11] KOCZURA R., MOKRACKA J., JABŁOŃSKA L., *et al.*, "Antimicrobial resistance of integron-harboring *Escherichia coli* isolates from clinical samples, wastewater treatment plant and river water", *Science of the Total Environment*, vol. 14, no. 1, 680–685, 2011.

[KOH 10] KOHANSKI M.A., DEPRISTO M.A., COLLINS J.J., "Sublethal antibiotic treatment leads to multidrug resistance via radical-induced mutagenesis", *Molecular Cell*, vol. 37, no. 3, pp. 311–320, 2010.

[KRI 11] KRISTIANSSON E., FICK J., JANZON A., *et al.*, "Pyrosequencing of antibiotic-contaminated river sediments reveals high levels of resistance and gene transfer elements", *PLoS One*, vol. 6, no. 2, p. e17038, 2011.

[KUM 09] KÜMMERER K., "Antibiotics in the aquatic environment – a review – Part I", *Chemosphere*, vol. 75, no. 4, pp. 417–434, 2009.

[LAN 09] LÄNGIN A., ALEXY R., KÖNIG A., *et al.*, "Deactivation and transformation products in biodegradability testing of ß-lactams amoxicillin and piperacillin", *Chemosphere*, vol. 75, no. 3, pp. 347–354, 2009.

[LAR 10] LAROCHE-AJZENBERG E., Etude de l'antibiorésistance de la population d'*Escherichia coli* isolée d'environnements aquatiques: estuaire et hydrosytèmes karstiques, Doctoral thesis, University of Rouen, 2010.

[LAR 09] LAROCHE E., PAWLAK B., BERTHE T., *et al.*, "Occurrence of antibiotic resistance and class 1, 2 and 3 integrons in *Escherichia coli* isolated from a densely populated estuary (Seine, France)", *FEMS Microbiology Ecology*, vol. 68, no. 1, pp. 118–130, 2009.

[LAR 07] LARSSON D.G., DE PEDRO C., PAXEUS N., "Effluent from drug manufactures contains extremely high levels of pharmaceutical", *Journal of Hazardous Materials*, vol. 148, no. 3, pp. 751–755, 2007.

[LEC 13] LECLERCQ R., OBERLÉ K., GALOPIN S., *et al.*, "Changes in Enterococcal populations and related antibiotic resistance along a medical center-wastewater treatment plant-river continuum", *Applied Environmental Microbiology*, vol. 79, pp. 2428–2434, 2013.

[LEA 06] LEAVIS H.L., WILLEMS R.J.L., TOP J., *et al.*, "High-level ciprofloxacin resistance from point mutations in *gyrA* and *parC* confined to global hospital-adapted clonal lineage CC17 of *Enterococcus faecium*", *Journal of Clinical Microbiology*, vol. 44, no. 3, pp. 1059–1064, 2006.

[LIN 04] LINDBERG R., JARNHEIMER P.A., OLSEN B., *et al.*, "Determination of antibiotic substances in hospital sewage water using solid phase extraction and liquid chromatography/mass spectrometry and group analogue internal standards", *Chemosphere*, vol. 57, no. 10, pp. 1479–1488, 2004.

[LUI 01] LUIS CAMPOS J., GARRIDO J.M., MENDEZ R., *et al.*, "Effect of two broad-spectrum antibiotics on activity and stability of continuous nitrifying system", *Applied Biochemistry and Biotechnology*, vol. 95, no. 1, pp. 1–10, 2001.

[MAR 08] MARTINEZ J.L., "Antibiotics and antibiotic resistance genes in natural environments", *Science*, vol. 321, pp. 365–367, 2008.

[MAR 09a] MARTINEZ J.L., "Environmental pollution by antibiotics and by antibiotic resistance determinants", *Environmental Pollution*, vol. 157, no. 11, pp. 2893–2902, 2009.

[MAR 09b] MARTINEZ J.L., "The role of natural environments in the evolution of resistance traits in pathogenic bacteria", *Proceedings of the Royal Society B: Biological Sciences*, vol. 276, no. 1667, pp. 2521–2530, 2009.

[MAR 06] MARUYAMA F., TANI K., KENZAKA T., *et al.*, "Quantitative determination of free-DNA uptake in river bacteria at the single-cell level by *in situ* rolling-circle amplification", *Applied and Environmental Microbiology*, vol. 72, no. 9, pp. 6248–6256, 2006.

[MAS 10] MASSEY L.B., HAGGARD B.E., GALLOWAY J.M., *et al.*, "Antibiotic fate and transport in three effluent-dominated Ozark streams", *Ecological Engineering*, vol. 36, no. 7, pp. 930–938, 2010.

[MIA 04] MIAO X.S., BISHAY F., CHEN M., *et al.*, "Occurrence of antimicrobials in the final effluents of wastewater treatment plants in Canada", *Environmental Science & Technology*, vol. 38, no. 13, pp. 3533–3541, 2004.

[NWO 01] NWOSU V.C., "Antibiotic resistance with particular reference to soil microorganisms", *Research in Microbiology*, vol. 152, pp. 421–430, 2001.

[OBE 12] OBERLÉ K., CAPDEVILLE M.J., BERTHE T., *et al.*, "Evidence for a complex relationship between antibiotics and antibiotic-resistant *Escherichia coli*: from medical center patients to a receiving environment", *Environmental Sciences Technology*, vol. 46, pp. 1859–1868, 2012.

[OMS 01] ORGANISATION MONDIALE DE LA SANTÉ, WHO Global strategy for containment of Antimicrobial resistance, 2001.

[PAU 91] PAU J.H., FRISCHER M.E., THURMOND J.M., "Gene transfer in marine water column and sediment microcosms by natural plasmid transformation", *Applied and Environmental Microbiology*, vol. 57, no. 5, pp. 1509–1515, 1991.

[PEI 06] PEI R., KIM S.C., CARLSON K.H., *et al.*, "Effect of river landscape on the sediment concentrations of antibiotics and corresponding antibiotic resistance genes (ARG)", *Water Research*, vol. 40, no. 12, pp. 2427–2435, 2006.

[PHA 13] PHAN K., FERENCI T., "A design-constraint trade-off underpins the diversity in ecologically important traits in species *Escherichia coli*", *ISME Journal*, vol. 7, pp. 2034–2043, 2013.

[POI 02] POIREL L., KÄMPFER P., NORDMANN P., "Chromosome-encoded Ambler class A beta-lactamase of *Kluyvera georgiana*, a probable progenitor of a subgroup of CTX-M extended-spectrum beta-lactamases", *Antimicrobial Agents Chemotherapy*, vol. 46, pp. 4038–4040, 2002.

[RAT 10] RATAJCZAK M., LAROCHE E., BERTHE T., *et al.*, "Influence of hydrological conditions on the *Escherichia coli* population structure in the water of a creek on a rural watershed", *BMC Microbiology*, vol. 10, p. 222, 2010.

[SAN 10] SANTOS L.H., ARAÚJO A.N., FACHINI A., *et al.*, "Ecotoxicological aspects related to the presence of pharmaceuticals in the aquatic environment", *Journal of Hazardous Materials*, vol. 175, nos. 1–3, pp. 45–95, 2010.

[SER 07] SERVAIS P., GARCIA-ARMISEN T., GEORGE I., *et al.*, "Fecal bacteria in the rivers of the Seine drainage network (France): sources, fate and modeling", *Science of the Total Environment*, vol. 375, no. 1, pp. 152–167, 2007.

[SEV 02] SÉVENO N., KALLIFIDAS D., SMALLA K., *et al.*, "Occurrence and reservoirs of antibiotic resistance genes in the environment", *Reviews in Medical Microbiology*, vol. 13, p. 15, 2002.

[STA 12] STALDER T., BARRAUD O., CASELLAS M., *et al.*, "Integron involvement in environmental spread of antibiotic resistance", *Frontiers in Microbiology*, vol. 3, p. 119, 2012.

[TAM 08] TAMTAM F., MERCIER F., LE BOT B., *et al.*, "Occurrence and fate of antibiotics in the Seine River in various hydrological conditions", *Science of the Total Environment*, vol. 393, no. 1, pp. 84–95, 2008.

[TAM 11] TAMTAM F., LE BOT B., DINH T., *et al.*, "A 50-year record of quinolone and sulphonamide antimicrobial agents in Seine river sediments", *Journal of Soils and Sediments*, vol. 11, pp. 852–859, 2011.

[TON 11] TONG L., EICHHORN P., PEREZ S., *et al.*, "Photodegradation of azithromycin in various aqueous systems under simulated and natural solar radiation: kinetics and identification of photoproducts", *Chemosphere*, vol. 83, no. 3, pp. 340–348, 2011.

[TOU 07] TOURON A., BERTHE T., GARGALA G., *et al.*, "Assessment of faecal contamination and the relationship between pathogens and faecal bacterial indicators in an estuarine environment (Seine, France)", *Marine Pollution Bulletin*, vol. 54, no. 9, pp. 1441–1450, 2007.

[WAL 11] WALSH T., BURKITT M., WELC C., *et al.*, "*Kluyvera ascorbata* bacteremia and meningitis: a case report and review of the literature", *Infectious Diseases in Clinical Practice*, vol. 19, no. 4, pp. 247–241, 2011.

[WAT 07] WATKINSON A.J., MICALIZZI G.B., GRAHAM G.M., *et al.*, "Antibiotic resistant *Escherichia coli* in wastewaters, surface waters and oysters from an urban riverine system", *Applied and Environmental Microbiology*, 2007.

[WAT 09] WATKINSON A.J., MURBY E.J., KOLPIN D.W., *et al.*, "The occurrence of antibiotics in an urban watershed: from wastewater to drinking water", *Science of the Total Environment*, vol. 407, no. 8, pp. 2711–2723, 2009.

[WEL 13] WELLINGTON E., BOXALL A.B., CROSS P., *et al.*, "The role of the natural environment in the emergent of antibiotic resistance in gram negative bacteria", *Lancet Infectious Disease*, vol. 13, pp. 155–165, 2013.

[WIS 02] WISE R., "Antimicrobial resistance: priorities for action", *Journal of Antimicrobial Chemotherapy*, vol. 49, no. 4, pp. 585–586, 2002.

[WRI 07] WRIGHT G.D., "The antibiotic resistome: the nexus of chemical and genetic diversity", *Nature Reviews Microbiology*, vol. 5, no. 3, pp. 175–186, 2007.

[WUN 11] WUNDER D.B., BOSSCHER V.A., COK R.C., *et al.*, "Sorption of antibiotics to biofilm", *Water Research*, vol. 45, no. 6, pp. 2270–2280, 2011.

[XU 09] XU W.H., ZHANG G., WAI O.W.H., *et al.*, "Transport and adsorption of antibiotics by marine sediments in a dynamic environment", *Journal of Soils and Sediments*, vol. 9, no. 4, pp. 364–373, 2009.

[YAN 13] YAN C., DINH Q.T., CHEVREUIL M., *et al.*, "The effect of environmental and therapeutic concentrations of antibiotics on nitrate reduction rates in river sediment", *Water Research*, vol. 47, pp. 3654–3662, 2013.

[ZHA 13] ZHANJUN L., WELSHAN W., YING Z., *et al.*, "Antibiotics at subinhibitory concentrarions improve the quorum sensing behavior of *Chromobacterium violaceum*", *FEMS Microbiological Letters*, vol. 341, pp. 37–44, 2013.

3

Microbiological Coastal Risks and Monitoring Systems

3.1. Introduction

Coastal zones have been inhabited by people for a long time, and today the majority of towns are situated in coastal zones. Marine-coastal waters receive enormous quantities of water, both used and polluted through industrial and agricultural activities. Local environmental dynamics can become global as the phenomena resulting from climate change show. As we will see later in this chapter, the physico-chemical modifications of parameters, whether they are due to the evolution of the climate or to the behavior of people toward their environment, can have consequences on the presence and development of pathogenic agents (viruses, bacteria and parasites) or their vectors in aquatic environments, on a regional as well as on a global scale [BAU 04].

The marine environment contains numerous indigenous microorganisms, including pathogens for marine organisms. Some of these microorganisms can also be pathogenic for humans. Other microorganisms pathogenic to humans are excreted in feces and are introduced into marine environments through treated or untreated waste water emissions.

Chapter written by Patrick MONFORT, Serge MORAND and Murielle LAFAYE.

The globalization of human population dynamics, economic growth, commercial exchanges, international tourism and the exploitation of marine resources leads to consequences, themselves global, for the coastal environment.

These modifications of the epidemiological environment can have important consequences for human health [BOW 08]. Researchers strive to develop models that, integrated into early warning monitoring systems, will help to develop and control public regulations on water quality (directive-framework on water and directive-framework strategy for the marine environment) and health.

We will undertake identification of environmental and climatic parameters favorable to the modification of the epidemiological environment; what pathogenic agents are favored in aquatic environments and what their causes are; what the impacts on the question of human health are; what environmental parameters are to be measured; how to reduce exposure and transmission to the host; how to develop monitoring and early warning systems.

3.2. Risks and infectious diseases linked to coastal regions

Aquatic environments are particularly favorable to the transmission of numerous infectious agents whether this is directly through the water (*water-borne*), often via an oral–fecal-type transmission, or through contamination/infection of food (*food-borne*), through the consumption of food harboring infectious agents, and finally through vectors that find optimal conditions there for habitation and reproduction (*vector-borne*), or vectorial transmission.

Coastal zones, characterized by the interface between sea and continent, notably estuaries, deltas and coastal lagoons, are favorable locations for the development of infectious diseases because of the densities of their human population, their significant biodiversity (which plays an important role for reservoirs and vectors) and activity linked to the water (agriculture, aquaculture, fishing, tourism and bathing).

3.2.1. *Pathogenic agents*

The aim of the following sections is not to cover the whole spectrum of infectious and parasitic maladies and their agents (viruses, bacteria, protozoa and helminthes), nor to describe their cycles, but to give some examples illustrating the principal modes of contamination; this is to better understand the potential impacts of global environmental changes on their epidemiological dynamics.

3.2.1.1. *Virus*

Numerous viruses pathogenic to humans are associated with gastroenteritis (norovirus, rotavirus, etc.), hepatitis (virus for hepatitis A or E), poliomyelitis and various infections (enterovirus) [BOS 05]. They are transmitted via the oral–fecal route through the water or contaminated food products. Contamination in humans through the marine environment essentially occurs through the consumption of shellfish following contamination of the marine environment by emissions from urban effluent. The survival of these viruses depends on different factors, such as temperature, solar radiation and adsorption. Even if these viruses are more or less broken down in the marine environment, the elevated quantities of the virus ejected permit the contamination of shellfish, which, through their significant capacity for filtration, will retain and concentrate the virus.

Other viruses are transmitted by mosquitoes, causing diseases such as dengue fever, chikungunya, Japanese encephalitis and West Nile Fever. The development of these mosquitoes requires concentrations of water for their larval stages, making the waters of deltas or coastal lagoons favorable locations for their proliferation, like continental waters [PON 07, RAM 12].

3.2.1.2. *Bacteria*

Pathogenic enteric bacteria are found in the marine environment following emissions of urban effluent, such as *Salmonella*, *Shigella* and certain *E. coli* enteropathogens, responsible for gastroenteritis of varying severity, or typhoid in the case of *Salmonella typhi*. The transmission is oral–fecal, and the consumption of shellfish is responsible for most contaminations [GIR 08]. The survival of enteric bacteria in marine environments has been studied in order to understand the molecular mechanisms at work [ROZ 05].

Other bacteria of non-fecal origin are also responsible for infections from food or serious illnesses, in particular bacteria of the *Vibrio* genus. These bacteria are indigenous to marine environments (they can cause death in aquacultural production, like that of oysters). More seriously, the presence of the species *Vibrio cholerae*, the cholera agent, can decimate human populations in countries where hygiene is poorly controlled. Other species of *Vibrio* cause diarrhea and skin infections and can be found varyingly concentrated in shellfish [PRU 05].

3.2.1.3. *Parasites*

Protists (unicellular organisms), which cause parasitic diseases, can be transmitted by arthropod vectors, or by the contamination of drinking water or food.

Malaria is caused by the parasite *Plasmodium*, and remains the foremost vectorial infectious disease in the world, principally in developing countries. The parasite is transmitted by vectors, mosquitoes of the *Anopheles* genus that need water points to ensure their development. Certain anopheles tolerate and develop in brackish waters (of variable salinity), which explains a widespread distribution of vectors for malaria even in coastal zones. The Camargue, the coastal lagoons of littoral Languedoc or the lagoon of Venice were, in past centuries, environments where malaria was rife. Malaria was, therefore, present up to the time of the Second World War in the Breton marshes and the Camargue [CAL 47]. In Asia, the coastal lagoons are also favorable environments for the development of mosquitoes that are vectors of malarial agents [RAM 12].

The protozoa *Giardia* is the main cause of diarrhea parasites in the world and can reach a prevalence of close to 50% in come tropical countries. Its transmission is oral–fecal, through cysts (a form of environmental resistance) present in water or food or by dissemination in vegetative form in used water. The protozoa of the genus *Cryptosporidium* are agents of diarrheas, sometimes accompanied by severe complications. The transmission is also oral–fecal (oocysts) through the ordure of wild and domestic animals. A large number of domestic and wild animals are reservoirs for these zoonotic protozoa (animal pathogen agents transmittable to humans). *Giardia* cysts and *Cryptosporidium* oocytes have been found in oysters and other shellfish, in particular after strong rains in drainage basins although for the moment, it has not been clearly established that infections occur through the consumption of shellfish. However, it seems that such infections are currently underestimated, as well as the risk linked to the consumption of raw or slightly cooked seafood [GIR 08].

Several zoonoses concern macroparasites of the helminthes type (cestodes, nematodes and trematodes). The transmission of these parasites can occur, for some species, through the consumption of intermediate infected hosts living in marine environments [GIR 08]. Thus, the consumption of fish from brackish waters, such as mullet or crabs and shrimps, is known to transmit certain trematodes (flukes) responsible for distomatosis. *Anisakis* is a type of nematode that infects most sea fish. It is responsible for anisakiase, caused by the consumption of raw (sushi, sashimi and ceviche) or half-cooked fish, and can cause anaphylactic shock or bowel obstruction, depending on the sensitivity of the individuals. European legislation obliges raw or partially cooked fish to be frozen for 24 h before consumption in order to kill the parasites present, which has enabled the incidence of parasite infections to be reduced.

3.2.1.4. *Toxic microorganisms*

More than 150 species of microalgae, including dinoflagellates, diatoms, raphidophytes, prymnesiophytes, pelagophytes and cyanobacteria, are harmful or toxic microorganisms present in coastal waters and interface systems (estuaries, delta and coastal lagoons). These toxinogenic microorganisms produce different toxins that can cause widespread mortality in marine animals, or accumulate (bioaccumulation) in the tissues of aquatic organisms, such as fish and shellfish, therefore, making them dangerous for human consumption [LAN 05].

Different thermoresistant neurotoxins are responsible for poisoning, which can be very serious: *Amnesic Shellfish Poisoning* (ASP) toxins, *Diarrheic Shellfish Poisoning* (DSP) toxins and *Paralytic Shellfish Poisoning* (PSP) toxins. Depending on the toxins, the symptoms are intestinal pains, diarrhea, vomiting, paralysis and comas. The most serious forms of poisoning are responsible for neurotoxic shock and death.

3.2.1.5. *Significance of infectious diseases linked to coastal environments for health*

Infections in food transmitted by the consumption of sea food in particular shellfish, are a significant cause of death [GIR 08].

The *disability-adjusted life year* (DALY) is an index developed in view of comparing the impact of diseases between themselves (and between countries) on human health. This index estimates the number of years of life spent incapacitated following an illness, as the sum of years potentially lost through premature death and number of productive years of life lost due to temporary or permanent incapacity.

Table 3.1 summarizes the data for some widespread infectious diseases. Diarrhea is the main cause of DALY and loss of human life, overwhelmingly among children in developing countries.

Infectious diseases	DALY (millions of days per year)	Percentage of DALY from a total of infectious illnesses, births, neonatal complications and food
Diarrhea	73	4.8
Malaria	34	2.3
Intestinal infections from nematodes	4	0.3
Dengue	0.67	0
VIH/SIDA	58	4.1

Table 3.1. *DALY of some infectious maladies that can be linked to humid coastal zones or marine areas. The case of HIV/AIDS is given for comparison (OMS, statistics from 2004, www.who.int)*

These diseases include those in which infections are linked to coastal environments, whether the transmission is hydric, vectorial or *via* food. The available data do not give the proportion of DALY caused by these infections, since very often there are no epidemiological studies refined enough to determine the cause of diarrhea or infections following consumption of sea produce or contact with water.

Some studies have enabled the DALY of infections linked to marine environments polluted by used water to be estimated [SHU 05]. These studies estimate a DALY of 180,000 days per year caused by respiratory infections (Ear, Nose and Throat (ENT)) and the gastroenteritis associated with bathing in seawater of acceptable quality (European and US norms). This estimation shows the low rate of infections by microorganisms of fecal origin transmitted during recreational activities in contaminated seawater.

The results of the estimation for hepatitis A and E transmitted through the consumption of raw shellfish give a DALY of 2.9 million days per year. This approach, even if it remains approximate, shows the significance of infectious diseases transmitted by marine environments. It certainly underestimates the reality of these diseases, since it does not take account of other pathogenic agents, such as viral gastroenteritis and vibriosis (apart from cholera), transmitted through the consumption of shellfish and raw or slightly cooked sea produce.

3.2.2. *Environmental change and modifications in the epidemiological environment*

3.2.2.1. *Climate change*

Climate change affects the distribution and incidence of infectious diseases through the effects of different modifications of environmental factors. Thus, the increase in air or water temperature favors the dispersion of pathogens or vectors from tropical zones to temperate zones. This is the case with bacteria of the *Vibrio* genus, pathogens in humans and aquatic animals, or certain mosquito vectors of malarial agents, dengue fever and chikungunya [GIT 01, VEZ 13].

Climate change also carries the consequence of increasing sea levels through the melting of ice on land (Greenland and the Antarctic) or through the thermal expansion of marine waters. This rise in water level causes a salinization of coastal continental waters (saline intrusion in estuaries, deltas, coastal lagoons or in aquifers). Several species of mosquitoes that are vectors of malarial agents, dengue or chikungunya, show a tolerance to changes in salinity and adaptability to its increase [RAM 12]. The risks of emergence, re-emergence or epidemics are, therefore, to be taken into consideration in all the deltas and coastal lagoons where these vectors have the chance to become established.

Climatic variability is a factor in the rise of epidemics. El Niño/La Niña episodes, as well as the *North Atlantic Oscillation* (NAO) variability, are linked to a rise in epidemics of many oral–fecal diseases (viral or microbial) or to vector transmission [MOR 12, MOR 13].

3.2.2.2. *Changes in biodiversity*

Global environmental changes also concern changes in biodiversity: drops in biodiversity (due to extinctions), migrations (climate change) and invasions (commerce).

Several studies suggest that the drop in biodiversity, experienced by a number of species, favors the transmission of vector diseases. Thus, epidemics of *West Nile Fever* seem to be encouraged by a drop in the number of avian species. An increased bird population could have the consequence of diluting mosquito bites among birds that do not act as reservoirs of the pathogenic agents that they carry, thus diminishing the rate of infected bites from hosts to these agents [KEE 10].

With the movement of 3–6 billion tonnes of ballast water, maritime transport is considered as the principal source of exchanges and invasions of species in coastal waters, with more than 10,000 species transported each year. Among them, microorganisms (pathogens and toxin genes), parasites, vectors and reservoir hosts constitute so many new risks for human health (and for the health of domestic and farm animals) [DRA 04, LAN 05]. Important treatment programs for ballast waters are being developed or put in place in developed countries (treatment with ultraviolet (UV) or electrolysis).

3.2.2.3. *Pollution and microalgae*

Intensification of agricultural activity and the use of fertilizer, nitrates and phosphates, as well as industrial pollution, impact continental waters and coastal marine waters by favoring the development of blooms of toxic microalgae [LAN 05]. The increase in these blooms has been reported in numerous estuaries and coastal lagoons with important consequences on the sanitary quality of water for swimming, and fishing and the production of shellfish.

3.2.3. *The consequences in terms of new infectious risks*

The quality of coastal waters, deltas, estuaries and coastal lagoons is a challenge in the face of demographic growth and the increasing number of towns on the coast, the intensification of agriculture and climate change. Efforts should be made to improve the quality of the environments (continental and coastal waters) and the quality of wastewater emissions, as well as the quality of drinking water. Monitoring of aquatic environments requires agricultural emissions to be taken into account as well as fluctuations in climate that influence the dynamic of infectious agents (or their production of toxins) and their vectors.

3.2.4. *Emergence: cases of human pathogenic vibrios*

3.2.4.1. *General remarks*

The *vibrio* genus belongs to the *Vibrionaceae* family and is composed of a hundred species. These bacteria are indigenous to marine environments and are present from marine-coastal environments with brackish waters as far as extreme oceanic depths, depending on the species [COL 77, PRU 05]. Vibrios are aero-anaerobic heterotrophs, which break down organic matter. They are thus involved in biogeochemical cycles and can represent a significant part of the total number of bacteria cultivable in coastal waters (up to 40%).

In addition to their capacity for biodegradation, certain species are pathogens for aquatic animals and humans. Numerous studies show that these pathogens are responsible for emergent epidemics, often related to climate change. If the rise in the temperature of marine waters favors these pathogens, the consequences of these climatic variations on the hosts also come into play to explain these emergences. Thus, the whitening of coral is triggered by global warming, which breaks these organisms down, and by *Vibrio shiloi* and *Vibrio coralliitycus* that are pathogenic for coral [ROS 05]. Intensive farming and anthropic pressure are also responsible for the development of serious epidemics that can decimate marine farms. This is the case with *Vibrio salmonicida*, responsible for significant epizootics of farmed salmon [AUS 05]. Other vibrios are mollusc pathogens, such as *Vibrio tapetis*, responsible for brown ring disease in clams, or *Vibrio splendidus* and *Vibrio aesturianus*, responsible for the death of oyster larvae, undoubtedly in association with a herpes virus [PAI 04, SCH 12].

V. cholerae is the most well known of the vibrios pathogenic for humans and is responsible for cholera. The majority of cases of non-choleric human vibriosis are due to *Vibrio parahaemolyticus*, *Vibrio vulnificus* and non-choleric *V. cholerae* [PRU 05]. These three vibrios are the subject of a detailed presentation in the following sections. Other vibrios are implicated in non-choleric vibriosis, such as *Vibrio alginolyticus*, *Vibrio damsela*, *Vibrio fluvialis*, *Vibrio furnissii*, *Vibrio hollisae*, *Vibrio metschnikovii* and *Vibrio mimicus*. In all these cases, the consumption of shellfish and seafood, raw or slightly cooked, is the cause of gastroenteritis from vibrios; the contact of wounds with seawater is responsible for skin infections that can evolve into septicemia.

3.2.4.2. *Vibrio cholerae, Vibrio parahaemolyticus, Vibrio vulnificus*

Cholera is a permanently endemic disease on the Indian subcontinent and in certain African and South American countries, with around 3 million cases per year, with a 2.4% mortality [WHO 11]. The disease is triggered by serogroups O1 and O139 of *V. cholerae*, which possess the choleric toxin (CT).

The non-choleric human pathogenic vibrios are responsible for numerous cases of infections following the consumption of seafood in Asian countries and the United States [DAN 00, SU 07].

V. cholerae non-O1/non-O139, present in estuary systems, is capable of causing gastroenteritis of varying severity after the consumption of shellfish.

Present in coastal waters, *V. parahaemolyticus* was identified as a pathogen for humans in 1950 in Japan and Asia, following infections originating from food [SU 07]. Since then, significant epidemics of gastroenteritis from *V. parahaemolyticus* have occurred on all continents, and their principal cause is the consumption of shellfish and raw or slightly cooked seafood products, which makes this bacteria one of the main pathogens, responsible not only for gastroenteritis, but also for the infection of wounds and septicemia.

Vibrio vulnificus, present in environments with variable salinity such as estuaries and coastal lagoons, is responsible for septicemia after the ingestion of shellfish by individuals who are weakened or who have immunodeficiency [OLI 06]. In the United States, *V. vulnificus* is responsible for 95% of all deaths linked to the consumption of sea produce. Moreover, infections due to these bacteria, following consumption of shellfish (essentially oysters), have the highest mortality rate (50–60%). *V. vulnificus* is also associated with the serious infection of wounds, causing gangrene.

The presence of non-choleric human pathogenic vibrios is already well documented for coastal waters and shellfish in European countries, in Italy, Spain, Denmark and Norway [BAR 99, BAU 06, HOI 98, MAR 08]. These vibrios have been isolated in waters and in shellfish along the entire coast of France (the Channel, the Atlantic and the Mediterranean) [CAN 13, DET 10, HER 02]. Some epidemics have been described in this country, although they are rarely severe

[GEN 00]. In France, a hundred cases of infection by *V. parahaemolyticus* were reported in 2001, following the consumption of mussels imported from Ireland [HER 05]. Since then, only sporadic cases of infection by *V. parahaemolyticus* have been reported [QUI 05].

3.2.4.3. *Ecology of the human Vibrio pathogens*

The work of Rita Colwell has revealed the presence of *V. cholerae* O1 in the waters of Chesapeake Bay, on the east coast of the United States, although there has been no cholera epidemic in the United States for a long time [COL 81]. Numerous studies have since enabled the understanding of the ecology of *V. cholerae*, which is a bacterium native to coastal zones, in particular to estuaries and deltas. Temperature, salinity, phytoplankton and copepods are major factors that explain the dynamic of *V. cholerae*. This understanding is necessary to explain the epidemiology of the disease, which was only considered in the light of human transmission during cholera epidemics [COL 05].

Notably, these studies have shown that blooms of phytoplankton lead to significant blooms of zooplankton, including copepods, which are known to be a natural host for *V. cholerae*. In effect, *V. cholerae* forms biofilms on the carapaces of copepods, in a symbiotic or commensal relationship, which thus becomes an important reservoir for these bacteria. The marshes cause the intrusion inland of coastal waters filled with zooplankton in large delta zones like that of Bangladesh, thus placing human populations and *V. cholerae* in contact. This has enabled some triggers for cholera epidemics to be identified [COL 05, VEZ 10].

The spatio-temporal distributions of non-choleric human pathogenic vibrios have been linked to environmental factors [JOH 12]. Temperature is the most important factor because it explains the seasonal distributions in the coastal waters of temperate zones, with maximal abundances in summer and autumn. Salinity is also a major factor of the dynamic of the vibrios. A decrease in salinity favors growth and proliferation in brackish systems at variable salinity. The phytoplankton and zooplankton explain the dynamic of the vibrios.

Climatic anomalies, in particular the increase in the sea surface temperature (SST), a consequence of climate change, have been linked to the increase in the presence of *V. parahaemolyticus* and *V. vulnificus* and the risks of infection associated with these bacteria [VEZ 13]. Thus, cases of gastroenteritis from *V. parahaemolyticus*, following the consumption of oysters in Washington State and Alaska in the United States, were triggered by an increase in SST along the Pacific Coast in 2004 [MAR 10]. This has also been shown to be the case with the epidemics of *V. parahaemolyticus* in Peru in 1991 linked to El Niño. In 1999, in Galicia (Spain), an epidemic of *V. parahaemolyticus*, following consumption of oysters, was reported along with an abnormal increase in SST along the coasts of Galicia [BAK 10]. Baker *et al.* [BAK 10] conclude their analysis of epidemics in the world and in Europe by drawing particular attention to an increase of cases in Europe over the coming years due to the warming of marine waters caused by climate change.

3.3. Monitoring of key environmental parameters

Specialists make a significant distinction between "continental waters", "littoral waters", "sea" and "ocean", those responsible for the environment need specific indicators on these different zones to respond to the need for public policy. But with a single change found in one of these zones impacting on the others, from now on, it is necessary to be in a position to anticipate the monitoring of these zones in a more systematic manner. This is why those responsible for the environment, considering these zones to be part of a continuum, need coherent information from a local to a regional or indeed national level, to monitor and administer this "integrated global system".

In the case of waterborne infectious diseases, certain environmental and climatic parameters can be associated with the development and proliferation of infectious agents or toxinogens or vectors (Table 3.2).

Agents or vectors	To monitor
Cyanobacteria	Water temperature Water color Density of algae
Vibrios	Water temperature Salinity Turbidity Chlorophyll *a* Coastal occupation zone (urbanization, bathing, etc.) Marine currents
Water snails (schistosomiase)	Cartography of zones in water Cartography of plant species Time of submersion in the zones (bathymetry)
Algae	Water temperature pH Dissolved oxygen Sea currents Dynamic of algal efflorescences

Table 3.2. *Parameters to monitor in relation to pathogenic or toxinogenic agents, or vectors for waterborne infectious diseases*

These parameters can be measured and monitored in different ways, which will be detailed below.

3.3.1. *Systems of measurement in situ*

These systems enable an instantaneous measurement to be made, at an instant *t*, of the physico-chemical parameters.

3.3.1.1. *Methods of measurement in situ*

Several types of equipment exist that enable measurement of the physico-chemical parameters of the water (Table 3.3).

Method of measurement	Parameters measured
Drifting buoys	Temperature Salinity (surface → 2,000 m in depth)
Boats	Temperature Chlorophyll *a* (surface → bottom)
Gliders	Temperature Salinity Marine currents Chlorophyll *a* Oxygen Suspended matter (surface →1,000 m)
Fixed buoys	Temperature Salinity Currents (surface) Essentials for the calibration of the models
Spectrometers mounted on ferries (regional)	Temperature Salinity Turbidity Chlorophyll *a* Suspended matter Oxygen pH Algae (types)
Gauges	Sea level

Table 3.3. *Measurements in situ, method of measurement and physico-chemical parameters measured*

3.3.1.2. *Advantages and limitations*

Measurements *in situ* permit instantaneous data to be obtained at a precise location and moment. They are "the" reference. Their limitation concerns the surface that they are able to cover; it is local, indeed regional. To cover a coastal strip demands substantial material and human involvement, as much for taking measurements as for their laboratory analysis. To obtain coverage on the scale of a sea or an ocean is very complicated and requires international collaborations and a harmonizing of measurement protocol.

Nevertheless, measurements *in situ* are indispensable to fit and validate the models.

3.3.2. *Measurement systems by satellite*

3.3.2.1. *Methods of measuring by satellite*

Since the 1990s, several satellite missions (Figure 3.1) have employed measuring instruments that permit global coverage for certain parameters (Table 3.4).

Instrument (type)	Parameters measured	Satellite	Instrument (name)
Altimeter	Height of the sea's surface Speed of surface wind Wave height Sea ice coverage	JASON-1 ENVISAT JASON-2	Poseidon-2 RA-2 Poseidon-3
Spectro radiometer	Chlorophyll *a* Organic matter Surface temperature Sea ice coverage	MODIS MERIS	Aqua ENVISAT
Infrared radiometer	Surface temperature	METOP ENVISAT Aqua Terra MeteoSat DMSP	AVHRR AATSR MODIS SEVIRI GOES
Radiometer microwaves	Atmospheric water vapor content Atmospheric liquid water content (clouds) Precipitation Concentration of sea ice, extent of coverage Surface temperature Salinity	DMSP TRMM Aqua ENVISAT JASON-1 JASON-2	SSM/I TMI AMSR-E MWR JMR AMR
Scaterometer	Speed of surface wind Rain Concentration of sea ice	METOP	ASCAT
Synthetic aperture radar (SAR)	Wind Waves Sea ice survey	RADARSAT-1 RADARSAT-2 ENVISAT	

Table 3.4. *Measurements by satellite, types of instruments, physico-chemical parameters measurements, satellites and name of the measuring instrument*

Figure 3.1. *Image of the A-Train carrying satellite AQUA which carries the sensor MODIS used to sense the color of the water (Source: NASA) (see color section)*

One pertinent use of satellites consists of considering the vector that contributes to the diffusion of a disease or family of given illnesses. Thus, it is possible, relying on the expertise of entomologists and biologists, to identify the key parameters favoring its development, and finally those that are potentially observable due to satellites. There again, remote sensing expertise will ensure that the resolution and treatment are adapted to the phenomena and the dynamic to be observed. The current state of the art [LAF 12] enables us to draw up Table 3.5.

3.3.2.2. *Advantages and limitations*

From satellites traveling around the Earth, measurements can be taken at any point on the globe with the same coherence. Regional and even global indicators can be developed and used by scientific communities and/or professionals working at sea.

Disease	Pathogenic agents or toxinogens	Parameters to be observed	Satellite data	Satellite products with added value
Cholera Non-choleric vibriosis	*Vibrio cholerae* O1 and O139 *Vibrio cholerae* non-O1/non-O139, *Vibrio parahaemolyicus, Vibrio vulnificus*	Coastal zones	Optic and radar satellite images	Classification and dynamic of vegetation Urban and development classification
		Evolution of the environment		
		Evolution of the urban environment		
		Localization of bathing zones		
		Water temperature	Satellite measurements of the sea surface temperature (SST) and Chlorophyll *a*	Survey maps of the sea surface temperature (SST) Survey maps for Chlorophyll *a*
		Dynamic of chlorophyll		
		Turbidity	Satellite color of the water	Algorithm CDOM
		Salinity	Non-applicable	Non-applicable
Liver diseases Diseases of the nervous system Kidney diseases	Cyanobacteria	Water temperature	Satellite measurements of the sea surface temperature (SST)	Survey maps of the sea surface temperature (SST)
		Color change	Measurements of reflectance	Maps of the detection of color change
		Density of the algae	Measurements of reflectance	Maps of algal efflorescence

Table 3.5. *Adequacy of satellites for the observation of parameters relative to some diseases transmittable via marine waters in littoral zones*

Moreover, as the instruments onboard the satellites make continuous measurements, it is possible to use a long series of temporal measurements and to understand the seasonal, annual or pluriannual dynamic of the phenomena (that is to say El Niño). Satellites complement the systems for measuring *in situ*, and benefit from these measurements for their calibration.

Nevertheless, some limitations remain. Therefore, with radiometers sometimes having measures deteriorated on the land-sea transition, specific regional algorithms should be developed if we want to integrate these measurements into monitoring systems for coastal waters. Research-development studies on the instruments for measurement, as well as on the algorithms integrating these measurements in input data, are being carried out to overcome these limitations.

3.3.3. *Some ideas to bear in mind*

It is possible to measure a number of physico-chemical parameters of the water and the characteristics of the ocean due to measuring systems *in situ* and/or by instruments mounted on satellites. The accumulation of a long series of temporal measurements enables the dynamic of the evolution of these phenomena to be observed. Although the systems for measurement *in situ* guarantee "measured on data", it is difficult to develop them over large distances. Satellites enable measurements to be carried out locally, regionally and even over the entire Earth with the same overall coherence. In Europe, some operational information services integrating the two types of measurement are being developed. These services regularly provide information bulletins and/or, due to modeling software, map for predicting the evolution of parameters linked to water quality or monitoring of the ocean.

For example, we cite:

– MARCoast (EU): www.marcoast.eu;

– MyOcean (EU): www.myocean.eu;

– PREVIMER (Fr): www.previmer.org;

– MERCATOR Océan (Fr): www.mercator-ocean.fr.

3.4. Toward remote monitoring systems and early warning systems applied to the remote monitoring of vibrios and algae

3.4.1. *Understanding the effect of global change*

3.4.1.1. *Vulnerability to global change and epidemics*

In 2001, Githeko *et al.* [GIT 01] wrote "according to current data, the interannual and interdecade variability of the climate could have a direct influence on the epidemiology of diseases with vectorial transmission, added to current factors favorable to these diseases, such as seasonal meteorological variations, the socio-economic situation, programs for vector control, modifications in the environment and resistance to medication". Climatic factors and the environmental modifications that flow from them are generally considered to be effective triggers for some epidemics by amplifying the dynamic of transmission and diffusion of the pathogenic agents or vectors. In this context of global change, where the risks of epidemics, pandemics and the re-emergence of infectious diseases are increasing, there is, therefore, an urgent need to improve our knowledge of the dynamic relationships between the climate, environment and health.

3.4.1.2. *To adapt, is to predict*

The emergence and re-emergence of infectious diseases with a strong epidemic potential push agents of public health to adapt their policies for managing veterinary and human health. This adaptation requires new means of predicting risks to be put in place. In this context, the study of climate-dependent infectious diseases requires knowledge of the environmental factors favorable to the emergence and propagation of the disease. A first step for predicting risk consists of identifying the proliferation zones of pathogen agents and/or the vectors. This knowledge then permits risk zones for human and animal populations to be identified and modeled and the arrival of these diseases to be anticipated by putting in place early warning systems (SAP). The prediction of these exposure zones for hosts and pathogens or hosts and vectors of infectious diseases can then allow strategies for adaptation to be put in place in order to lessen the impact of global change.

3.4.2. *"Remote sensing" approach*

The *Centre National d'Etudes Spatiales* (CNES) (French Space Agency) has developed a concept that relies on a determinist approach to relationships between the climate, environment and health and on an original, tailored space package [CNE 08, LAC 07]. The conceptual approach is based on the study of the mechanisms favoring the appearance and propagation of infectious diseases by linking several disciplines such as environmental sciences, climate study, the biology of organisms, microbiology, ecology, etc.

As schematized below by Cécile Vignolles, responsible for the remote sensing program at CNES, this approach consists of three steps (Figure 3.2).

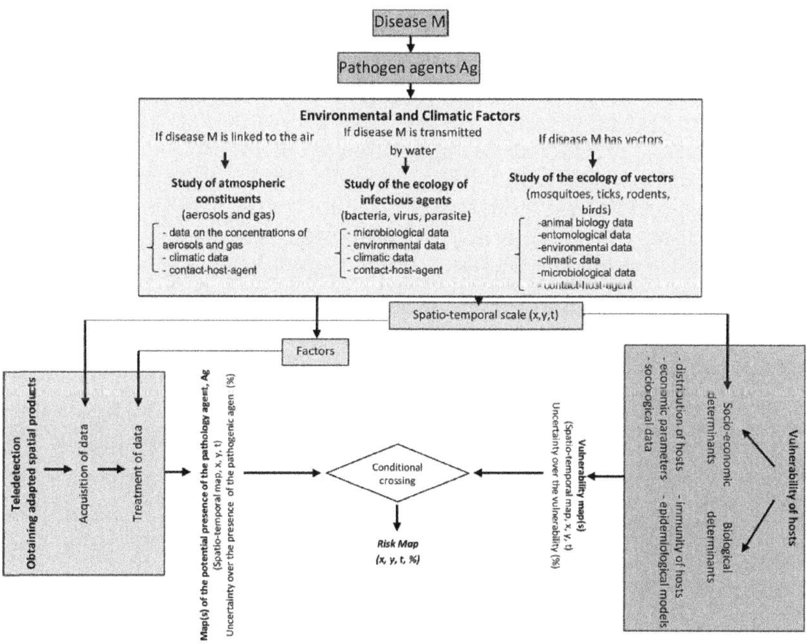

Figure 3.2. *Conceptual remote sensing approach by Cécile Vignolles, CNES 2008*

3.4.2.1. *Step 1: predict and understand the mechanisms for the emergence and propagation of the disease*

This first step consists of experimentally establishing the mechanisms that link the physical, biological and socio-economic parameters associated with the diseases. The objective is to identify the conditions favorable to the development of the disease and better to comprehend the mechanisms for the transmission and proliferation of pathogenic agents. This step involves an observation phase and a diagnostic phase:

– observation phase: its goal is to collect multidisciplinary data *in situ*: the ecology of the pathogen, the entomology of the vectors responsible for the pathogen's transmission, host–vector relationships or host–pathogen relationships, environmental and climatic data, sensitivity to the infectious agent, dose-infection, etc.;

– diagnostic phase: its goal is to extract and identify the principal mechanisms at work favoring the propagation and diffusion of the disease.

3.4.2.2. *Step 2: production of adapted space products*

This step should be taken in response to criteria specific to a disease, defined during the stage of understanding the processes governing this disease. Its objective is, therefore, to create and/or use appropriately adapted space technology. The satellite images provide information that does not directly concern the pathogens responsible for the disease, but rather their environment (geographic data, meteorological data, hydrological data, etc.) and notably the locations favoring their development and proliferation. The supply of satellite imagery is, therefore, to enable the measurement of the environmental factors favorable (or not) to the emergence of infectious diseases as identified in the course of the previous step: the identification of zones of vegetation, the detection of stagnant water, the surface temperature of the sea, the concentration of chlorophyll *a* (Chl *a*), salinity, etc., for example. This step consists of identifying which types of satellite data respond best to researchers' needs (spatio-temporal and spectral resolution) and then creating the products. The latter depend in general not only on new products, but also on the adaptation of already existing products; they are then validated by measurements of the terrain *in situ*, the advice of experts and potential users; the conception of an operational production chain can then be envisaged.

3.4.2.3. *Step 3: elaboration of predictive models*

This third step calls upon biomathematic modeling of epidemic dynamics (risk quantification), integrating the transport processes involved: pathogenic agents, vectors and hosts, physical and socio-ecological environments. This step notably involves the development of health information systems. The final objective is to put early warning systems in place (SAP), enabling epidemics to be predicted, either by creating such systems, or by integrating the biomathematical models developed in the existing systems.

The CNES currently applies this concept to several infectious diseases widespread in the world:

– Rift Valley fever in Senegal [GIC 13];

– rural and urban malaria on the African continent [MAC 11];

– dengue fever in Argentina and the Caribbean;

– *vibrio* (Vibriose) diseases around the Mediterranean basin;

– bilharzias in China.

3.4.3. *Application of remote monitoring to vibrios*

3.4.3.1. *The case of Chesapeake Bay, United States*

A system for predicting the probability of the presence of *V. cholerae* in Chesapeake Bay, United States, and also other pathogenic species of the same genus, such as *V. vulnificus*, has been developed [BAN 12, CON 09]. This system identifies and maps the geographical zones of the Bay where the environmental conditions are those of the bacterium's habitat. It was developed by using empirical models of the habitat in Chesapeake Bay for pathogenic species of the genus *vibrio* [JAC 10, LOU 03], and based on projections for the surface temperature of the water and salinity based on a model of the regional oceanic system (ROMS) configured for the Bay (ChesROMS).

ChesROMS is the free application (*open source*) of ROMS adapted for Chesapeake Bay. ROMS/ChesROMS is a model of primitive equations used by a broad community of scientists for numerous and diverse applications. ChesROMS uses historical data of reanalysis, observations in near real time and the prediction of data under external constraints, such as heat fluxes, river flows and even the ocean's sea level; the goal is to simulate the salinity and the surface temperature of the water used to generate predictions for the past, present and future of the probability of the presence of *Vibrio* [BAN 12, CON 09].

The short- and medium-term predictions of the probability of the presence of *V. cholerae*, *V. parahaemolyticus* and *V. vulnificus*, as shown in the maps in Figure 3.3, are generated using predictions of the surface temperature of the water and the salinity across habitat models [URQ 14]. For short-term predictions, the forcing factors were the same as those used for the reconstruction *a posteriori* of past predictions obtained from the North-American model on mesoscale and provided by the National Center for Environmental Prediction (NCEP).

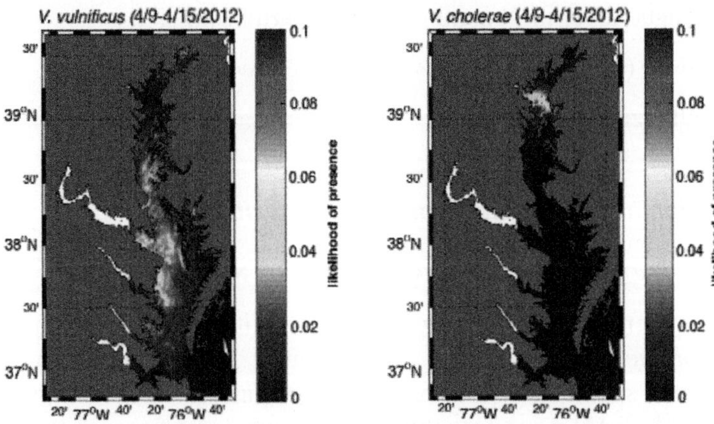

Figure 3.3. *Probability maps for the presence of V. vulnificus and V. cholerae in Chesapeake Bay (United States) generated using interpolated estimations of the surface temperature of the water and the salinity applied to models of the habitat of Vibrio, over the period from 9 to 15 April 2012. These maps have been kindly provided by E. Urquhart (Johns Hopkins University, Baltimore, United States) and R. Colwell (University of Maryland, United States) within the framework of collaboration (see color section)*

3.4.3.2. *The case of the Gulf of Bengal, India and Bangladesh*

A statistical model has been constructed relying on data on the dynamic of cholera in Calcutta (India) and Matlab (Bangladesh) over the period 1998–2006 [CON 08]. This model uses monthly data provided by remote sensing of the SST and Chl *a* in the Gulf of Bengal as well as pluviometry.

Because these data evolve seasonally, it is necessary to research these anomalies, that is to say the events that are not linked to seasonal cycles. These anomalies, calculated using averages for a given month and over the entire period (SST, Chl *a* and pluviometry) are removed from the data for each month. The authors also take account, in a provisional model of the cholera dynamic, of a latency period of one to two months between the variations of the two parameters (SST and Chl *a*) and the epidemiology of the disease in the population. One of the remarkable results of these studies shows that an increase of 1 mg/m^3 of the average anomaly of Chl *a* explains an increase of approximately 32% of cholera cases in Matlab with a latency time of a month.

By using such a predictive model on a regional scale (the Gulf of Bengal), it is possible to predict the health risks of cholera epidemics associated with climatic conditions, and above all, with their variations.

3.4.4. *Application of remote monitoring to cyanobacteria*

In Europe, cyanobacteria are the subject of particular monitoring (European directive 76160/CEE and 2006-7/CE) in order to reduce the population's exposure, which most often occurs when bathing. The bacteria's presence is locatable visually by a change in water color and the formation of filaments. This detection of change can be carried out due to satellite sensors dedicated to the observation of the water color (MERIS). In effect, the cyanobacteria have a spectral signature (Figure 3.4) measurable by remote sensing, enabling cyanobacteria blooms to be observed (Figure 3.5).

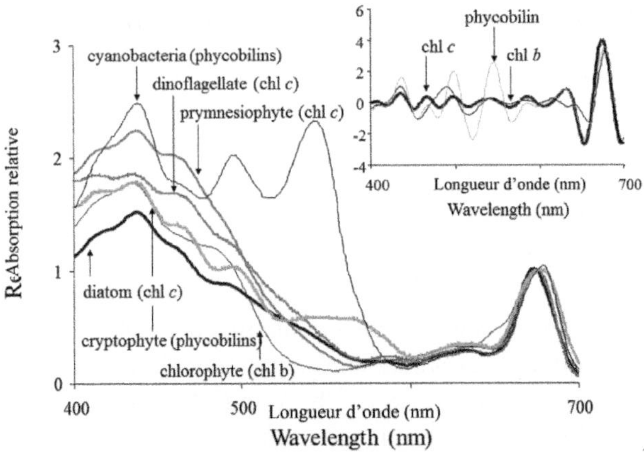

Figure 3.4. *Spectrum of relative absorption of different species of algae (including cyanobacteria), from Schofield, 2003 (overview of optical observation of biological variability, in HABWatch workshop, Villefranche-sur-Mer, 11–21 June 2003: www.obs-vlfr.fr/habwatch/)*

Figure 3.5. *Aerial view of a bloom of cyanobacteria in the Baltic Sea, 13 July 2005 (Source: MERIS-RR (ENVISAT-ESA); credits: ESA) (see color section)*

In the Baltic Sea, chronic episodes of cyanobacteria bloom occur each summer, between June and September (Figure 3.6). Within the framework of the Helsinki Convention, the neighboring countries have combined their scientific efforts to develop a system for predicting the cyanobacteria risk on a daily basis (Figure 3.7).

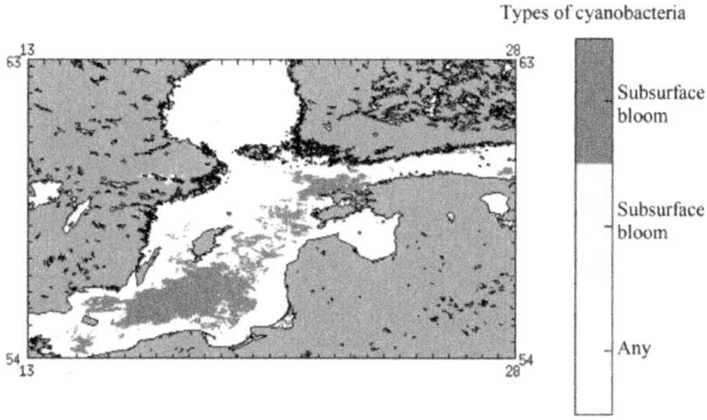

Figure 3.6. *Map of cyanobacteria bloom in the Baltic Sea for the day 11 July 2010 (sensor MERIS (ENVISAT-ESA), L3 MEGS, handling ACRI-ST) (see color section)*

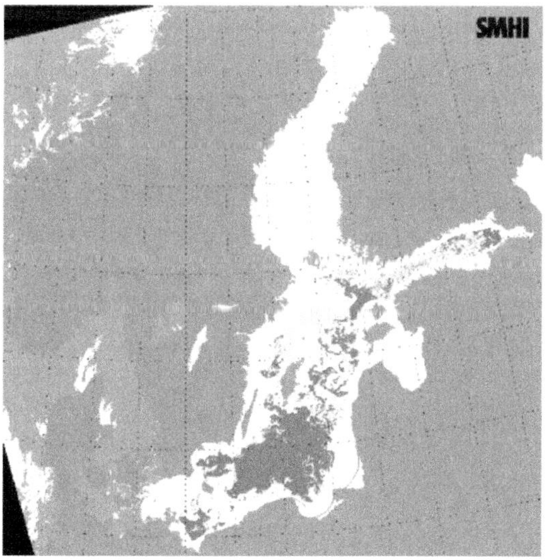

Figure 3.7. *HELCOM bulletin from 11 July 2010 (Source: Swedish Meteorological and Hydrological Institute – SMHI) (see color section)*

3.4.5. *Integration of models in the monitoring and alert systems*

The integration of the models in monitoring systems constitutes an aid for public policy on water management and health. Thus, this information can be integrated in the systems for aiding decisions and the management of environmental and sanitary risks. The maps for predicting the risk of the presence of pathogenic agents can be used to put in place the strategies for adaptation and plans for risk management.

Combined with predictions for climate evolution, these risk maps for pathogen presence will enable decision makers to make projections for the future: will the risk be elevated? Which zones will be affected? Which are those areas where a particular effort should be made to reduce the risk of an epidemic?

In themselves, however, these risk maps for the presence of vibrios or cyanobacteria are already useful information for populations and professionals living or working on the sea. Workers in aquaculture can use these predictions to optimize their production management with a view to bacterial risk throughout the year. Although news bulletins on water quality for bathing are already available for European citizens, making new predicative information available several days in advance could interest those wishing to engage in nautical activities.

In the United States, for example, the Maryland Healthy Beaches website enables citizens to view information on the evolution of bacterial risk and on the presence of blooms of algae on the beaches in Maryland (www.marylandhealthybeaches.com). This information site is supported by the Maryland Department of the Environment (MDE) and the Maryland Department of Health and Mental Hygiene with the help of the Environmental Protection Agency (EPA).

Another example in Europe concerns the monitoring of developments of cyanobacteria: due to bulletins delivered by the Swedish Meteorological and Hydrological Institute, the groups responsible for taking samples that can be directed as a priority to suspect sites to quantify the concentration of bacteria and decide if measures to restrict access to the affected zones should be put in place or not.

In this way, these prediction maps, initially developed within the context of research activity for a better comprehension of the phenomena and their dynamic, are gradually coming to be used in information systems and to aid operational decisions.

3.5. Acknowledgments

The authors thank Bernard Baleux for his reading and informed criticism of this chapter, as well as Sylvie Joussaume and Chantal Pacteau from GIS "Climat, Environnement, Société", for their help with the meetings and discussions that enabled this chapter to be written.

3.6. Bibliography

[AUS 05] AUSTIN B., "Bacterial pathogens of marine fish", in BELKIN S., COLWELL R.R. (eds.), *Oceans and Health: Pathogens in the Marine Environment*, Springer, New York, pp. 391–413, 2005.

[BAK 10] BAKER-AUSTIN C., STOCKLEY L., RANGDALE R., *et al.*, "Environmental occurrence and clinical impact of *Vibrio vulnificus* and *Vibrio parahaemolyticus*: a European perspective", *Environmental Microbiology Reports*, vol. 2, pp. 7–18, 2010

[BAN 12] BANAKAR V., CONSTANTIN DE MAGNY G., JACOBS J., *et al.*, "Temporal and spatial variability in the distribution of *Vibrio vulnificus* in the Chesapeake Bay: a hindcast study", *EcoHealth*, vol. 8, pp. 456–467, 2012.

[BAR 99] BARBIERI E., FALZANO L., FIORENTINI C., *et al.*, "Occurrence, diversity, and pathogenicity of halophilic *Vibrio* spp. and non-O1 *Vibrio cholerae* from estuarine waters along the Italian Adriatic coast", *Applied Environmental Microbiology*, vol. 65, pp. 2748–2753, 1999.

[BAU 04] BAUDA P., MONFORT P., "Agents pathogènes et modifications des environnements: quels risques actuels et futurs?", *Environnement, Risques et Santé*, vol. 3, pp. 165–172, 2004.

[BAU 06] BAUER A., OSTENSVIK O., FLORVAG M., *et al.*, "Occurrence of *Vibrio parahaemolyticus, V. cholerae*, and *V. vulnificus* in Norwegian blue mussels (*Mytilus edulis*)", *Applied Environmental Microbiology*, vol. 72, pp. 3058–3061, 2006.

[BOS 05] BOSCH A., ABAD F.X., PINTO R.M., "Human pathogenic viruses in the marine environment", in BELKIN S., COLWELL R.R. (eds.), *Oceans and Health: Pathogens in the Marine Environment*, Springer, New York, pp. 109–131, 2005.

[BOW 08] BOWEN R.E., "Globalization, and global ocean change: an overview of influences on human health", in WALSH P.J., SMITH S.L., FLEMING L.E. (eds.), *et al.*, *Oceans and Human Health, Risks and Remedies from the Seas*, Academic Press, pp. 19–23, 2008.

[CAL 47] CALLOT J., "Un problème complexe: la régression du paludisme en France", *Annales Economies, Sociétés, Civilisations*, vol. 2, pp. 328–335, 1947.

[CAN 13] CANTET F., HERVIO-HEATH D., CARO A., *et al.*, "Quantification of *Vibrio parahaemolyticus, V. vulnificus,* and *V. cholerae* in French Mediterranean coastal lagoons", *Research in Microbiology*, vol. 164, pp. 867–874, 2013.

[CNE 08] CNES, Method for tele-epidemiology (Méthode pour la télé-épidémiologie), Patent delivered, BFF 07P0650, 2008.

[COL 77] COLWELL R.R., KAPER J., JOSEPH S.W., "*Vibrio cholerae, Vibrio parahaemolyticus*, and other vibrios: occurrence and distribution in Chesapeake Bay", *Science*, vol. 198, pp. 394–396, 1977.

[COL 81] COLWELL R.R., SEIDLER R.J., KAPER J., *et al.*, "Occurrence of *Vibrio cholerae* serotype O1 in Maryland and Louisiana estuaries", *Applied and Environmental Microbiology*, vol. 41, pp. 555–558, 1981.

[COL 05] COLWELL R.R., "Global microbiology ecology of *Vibrio cholerae*", in BELKIN S., COLWELL R.R. (eds.), *Oceans and Health: Pathogens in the Marine Environment*, Springer, New York, pp. 297–305, 2005.

[CON 08] CONSTANTIN DE MAGNY G., MURTUGUDDE R., SAPIANO M.R.P., *et al.*, "Environmental signatures associated with cholera epidemics", *Proceedings of the National Academy of Sciences of the United States of America*, vol. 105, pp. 17676–17681, 2008.

[CON 09] CONSTANTIN DE MAGNY G., LONG W., BROWN C.W., *et al.*, "Predicting the distribution of *Vibrio* spp. in the Chesapeake Bay: a *Vibrio cholerae* case study", *EcoHealth*, vol. 6, pp. 378–389, 2009.

[DAN 00] DANIELS N.A., MACKINNON L., BISHOP R., *et al.*, "*Vibrio parahaemolyticus* infections in the United States, 1973-1998", *Journal of Infectious Diseases*, vol. 181, pp. 1661–1666, 2000.

[DET 10] DETER J., LOZACH S., VERON A., *et al.*, "Ecology of pathogenic and non-pathogenic *Vibrio parahaemolyticus* on the French Atlantic coast. Effects of temperature, salinity, turbidity and chlorophyll *a*", *Environmental Microbiology*, vol. 12, pp. 929–937, 2010.

[DRA 04] DRAKE J.M., LODGE D.M., "Global hot spots of biological invasions: evaluating options for ballast-water management", *Proceedings of the Royal Society London B*, vol. 271, no. 1539, pp. 575–580, 2004.

[GEN 00] GENESTE C., DAB W., CABANES P.A., *et al.*, "Les vibrioses non cholériques en France: cas identifiés de 1995 à 1998 par le Centre national de référence", *Bulletin épidémiologie hebdomadaire*, vol. 9, pp. 38–40, 2000.

[GIC 13] GICC 2013, Gestion et Impacts du Changement climatique, Programme de recherche, Appel à projets 2008, Résultats scientifiques et acquis pour les gestionnaires et les décideurs, ADAPTFVR – Impacts du changement climatique sur l'émergence des vecteurs de la Fièvre de la Vallée du Rift au Sénégal: adaptation et stratégie pour une meilleure gestion du pastoralisme au Sahel, pp. 61–79, 2013.

[GIR 08] GIRONES R., BOFILL-MAS S., FURONES D., *et al.*, "Foodborne infectious diseases and monitoring of marine food resources", in WALSH P.J., SMITH S.L., FLEMING L.E., *et al.*, (eds.), *Oceans and Human Health, Risks and Remedies from the Seas*, Academic Press, New York, pp. 359–380, 2008.

[GIT 01] GITHEKO A.K., LINDSAY S.W., CONFALONIERI U.E., *et al.*, "Changement climatique et maladies à transmission vectorielle: une analyse régionale", *Bulletin de l'Organisation mondiale de la Santé*, vol. 4, pp. 62–72, 2001.

[HER 02] HERVIO-HEATH D., COLWELL R.R., DERRIEN A., *et al.*, "Occurrence of pathogenic vibrios in coastal areas of France", *Journal of Applied Microbiology*, vol. 92, pp. 1123–1135, 2002.

[HER 05] HERVIO-HEATH D., ZIDANE M., LE SAUX J.C., *et al.*, "Toxi-infections alimentaires collectives liées à la consommation de moules contaminées par *Vibrio parahaemolyticus*: enquête environnementale", *Bulletin épidémiologique*, AFSSA, vol. 17, pp. 1–2, 2005.

[HOI 98] HOI L., LARSEN J.L., DALSGAARD I., *et al.*, "Occurrence of *Vibrio vulnificus* biotypes in Danish marine environments", *Applied Environmental Microbiology*, vol. 64, pp. 7–13, 1998.

[JAC 10] JACOBS J.M., RHODES M.R., BROWN C.W., *et al.*, "Predicting the distribution of *Vibrio vulnificus* in Chesapeake Bay", NOAA Technical Memorandum NOS NCCOS 112. NOAA National Centers for Coastal Ocean Science, Center for Coastal Environmental Health and Biomolecular Research, Cooperative Oxford Laboratory, Oxford, MD, p. 12, 2010.

[JOH 12] JOHNSON C.N., BOWERS J.C., GRIFFITT K.J., *et al.*, "Ecology of *Vibrio parahaemolyticus* and *Vibrio vulnificus* in the coastal and estuarine waters of Louisiana, Maryland, Mississippi, and Washington (United States)", *Applied Environmental Microbiology*, vol. 78, pp. 7249–7257, 2012.

[KEE 10] KEESING F., BELDEN L.K., DASZAK P., *et al.*, "Impacts of biodiversity on the emergence and transmission of infectious diseases", *Nature*, vol. 468, pp. 647–52, 2010.

[LAC 07] LACAUX J.P., TOURRE Y.M., VIGNOLLES C., *et al.*, "Classification of ponds from high-spatial resolution remote sensing: application to Rift Valley fever epidemics in Senegal", *Remote Sensing of Environment*, vol. 106, pp. 66–74, 2007.

[LAF 12] LAFAYE M., Etat de l'art en Environnement-Climat-Santé, Règlementation, besoins des acteurs, réponses des satellites et perspectives dans le domaine de l'eau, de l'air et des vecteurs en France, Note CNES, Réf.DSP/ARP/AV/2012.0002191, 2012.

[LAN 05] LANDSBERG J., VAN DOLA F., DOUCETTE G., "Marine and estuarine harmful algal blooms: impacts on human and animal health", in BELKIN S., COLWELL R.R. (eds.), *Oceans and Health: Pathogens in the Marine Environment*, Springer, New York, pp. 165–215, 2005.

[LOU 03] LOUIS V.R., RUSSEK-COHEN E., CHOOPUN N., *et al.*, "Predictability of *Vibrio cholerae* in Chesapeake Bay", *Applied Environmental Microbiology*, vol. 69, pp. 2773–2785, 2003.

[MAC 11] MACHAULT V., BORCHI F., VOUNATSOU P., *et al.*, "The use of remotely sensed environmental data in the study of malaria", *Geospatial Health*, vol. 5, pp. 151–168, 2011.

[MAR 08] MARTINEZ-URTAZA J., LOZANO-LEON A., VARELA-PET J., *et al.*, "Environmental determinants of the occurrence and distribution of *Vibrio parahaemolyticus* in the rias of Galicia, Spain", *Applied Environmental Microbiology*, vol. 74, pp. 265–274, 2008.

[MAR 10] MARTINEZ-URTAZA J., BOWERS J.C., TRINANES J., *et al.*, "Climate anomalies and the increasing risk of *Vibrio parahaemolyticus* and *Vibrio vulnificus* illness", *Food Research International*, vol. 43, pp. 1780–1790, 2010.

[MOR 12] MORAND S., WARET-SZKUTA A., "Les déterminants des maladies infectieuses humaines en Europe: influences de la biodiversité et de la variabilité climatique", *Bulletin épidémiologique hebdomadaire*, vol. 12–13, pp. 156–159, 2012.

[MOR 13] MORAND S., OWERS K., WARET-SZKUTA A., *et al.*, "Climate variability and outbreaks of infectious diseases in Europe", *Nature Scientific Reports*, vol. 3, p. 1774, 2013.

[OLI 06] OLIVER J.D., "*Vibrio vulnificus*", in THOMPSON F.L., AUSTIN B., SWINGS J. (eds.), *Biology of Vibrios*, ASM Press, Washington, pp. 349–366, 2006.

[PAI 04] PAILLARD C., "Review of brown ring disease, a vibriosis affecting clams", *Aquatic Living Resources*, vol. 17, pp. 467–475, 2004.

[PON 07] PONÇON N., BALENGHIEN T., TOTY C., *et al.*, "Effects of local anthropogenic changes on potential malaria vector *Anopheles hyrcanus* and West Nile Virus vector *Culex modestus*, Camargue, France", *Emerging Infectious Diseases*, vol. 13, no. 12, pp. 1810–1815, 2007.

[PRU 05] PRUZZO C., HUQ A., COLWELL R.R., *et al.*, "Pathogenic *Vibrio* species in the marine and estuarine environment", in BELKIN S., COLWELL R.R. (eds.), *Oceans and Health: Pathogens in the Marine Environment*, Springer, New York, pp. 217–252, 2005.

[QUI 05] QUILICI M.L., ROBERT-PILLOT A., PICART J., *et al.*, "Pandemic *Vibrio parahaemolyticus* O3:K6 spread, France", *Emerging Infectious Diseases Journal,* vol. 11, pp. 1148–1149, 2005.

[RAM 12] RAMASAMY R., SURENDRAN S.N., "Global climate change and its potential impact on disease transmission by salinity-tolerant mosquito vectors in coastal zones", *Frontiers in Physiology*, vol. 3, p. 198, 2012.

[ROS 05] ROSENBERG E., BARASH Y., "Microbial diseases of corals", in BELKIN S., COLWELL R.R. (eds.), *Oceans and Health: Pathogens in the Marine Environment*, Springer, New York, pp. 415–430, 2005.

[ROZ 05] ROZEN Y., BELKIN S., "Survival of enteric bacteria in seawater: molecular aspects", in BELKIN S., COLWELL R.R. (eds.), *Oceans and Health: Pathogens in the Marine Environment*, Springer, New York, pp. 93–107, 2005.

[SCH 03] SCHOFIELD O., "Overview of optical observation of biological variability", in *HABWatch workshop*, Villefranche-sur-Mer, pp. 11–21 June 2003. Available at: www.obs-vlfr.fr/habwatch/.

[SCH 12] SCHMITT P., DUPERTHUY M., MONTAGNANI C., *et al.*, "Immune responses in the Pacific oyster *Crassostrea gigas*: an overview with focus on summer mortalitie", in QIN J.G. (ed.), *Oysters: Physiology, Ecological Distribution and Mortality*, Nova Science Publishers, New York, 2012.

[SHU 05] SHUVAL H., "Thalassogenic infectious diseases caused by wastewater pollution of the marine environment: an estimate of worldwide occurrence", in BELKIN S., COLWELL R.R. (eds.), *Oceans and Health: Pathogens in the Marine Environment*, Springer, New York, pp. 373–389, 2005.

[SU 07] SU Y.C., LIU C., "*Vibrio parahaemolyticus*: a concern of seafood safety", *Food Microbiol*ogy, vol. 24, pp. 549–558, 2007.

[URQ 14] URQUHART E.A, ZAITCHIK B.F., WAUGH D.W., *et al.*, "Uncertainty in model predictions of *Vibrio vulnificus* response to climate variability and change: a Chesapeake Bay case study", *Plos One*, vol. 9, pp. 1–7, 2014.

[VEZ 10] VEZZULI L., PRUZZO C., HUQ A., *et al.*, "Environmental reservoirs of *Vibrio cholerae* and their role in cholera", *Environmental Microbiology Reports*, vol. 2, pp. 27–33, 2010.

[VEZ 13] VEZZULI L., COLWELL R.R., PRUZZO C., "Ocean warming and spread of pathogenic vibrios in the aquatic environment", *Microbial Ecology*, vol. 65, pp. 817–825, 2013.

[WHO 11] WHO, WORLD HEALTH ORGANIZATION, Weekly Epidemiological Record: Cholera 2010, vol. 32, pp. 325–340, 2011.

4

Vulnerability, Impacts and Adaptation of Coastal Zones to Global Change

4.1. Introduction

4.1.1. *The coastal zones*

Seashores are the land regions of the Earth contiguous to the sea, while coastal zones are the regions where the sea and land processes interact. They attract and are home to a large human population because of their important socio-economic value, and also contain a great variety of ecosystems rich in biodiversity. A coastal zone is the interface or transition zone between land and sea. For management purposes, various limits on the sea-side and on the land-side are used to define the coastal zone; these can be narrow and precise or wider and more nebulous. Here, the concept of low elevation coastal zone will often be used, defined as the land, population, infrastructure and ecosystems below 10 m above sea level [VAF 11]. The definition of the extension of the coastal zone on the land-side depends on the legal definition adopted, which takes account of political, social, economic and environmental considerations.

From a conceptual viewpoint, it is important to recognize that coastal zones constitute a complex adaptive system with two components: natural systems and human systems that interact with

---

Chapter written by Filipe DUARTE SANTOS.

each other. The natural coastal systems include different types of geological formation and ecosystem: rocky coasts, estuaries, barrier islands, beaches, sand dunes, estuaries, lagoons, deltas, river mouths, wetlands and coral reefs. Moreover, these systems support a large variety of services of a social, economic and cultural nature and are affected by various human activities. For human coastal systems, or coastal anthroposystems, it is possible to distinguish three different components: the first consists of the built environment, including settlements (residential and tourism facilities), the land and maritime transport infrastructures, industrial infrastructures and port facilities; the second comprises the human activities that take place on the coast, such as fishing, aquaculture, sport, leisure, tourism and activities that take place away from the coast but which exert pressure on the coast, such as the construction of dams, pollution of the soil and surface waters and deforestation; finally, the third component is the institutional, legislative, judicial and cultural activities that support the territorial organization and governing of coastal zones. Several recent studies have shown the advantages, for sustainable development, of considering the coastal zones as integrated ecological–social–economic systems [HOP 12, NEW 12]. Furthermore, it should be emphasized that coastal zone management must integrate the natural drivers of change and the anthropogenic actions that affect their dynamics.

4.1.2. *Global change*

Global environmental change refers to planetary-scale changes; in a broad sense it results from the evolution of the Earth system and the interactions and physical, chemical and biological processes in its subsystems – the lithosphere, the hydrosphere, the cryosphere, the atmosphere and the biosphere. Depending on their origin, natural and anthropogenic global changes can be distinguished; the identification and differentiation between the two is often difficult and subject to significant uncertainties. Many examples of natural global change can be found since the formation of the Earth around 4.5 billion years ago. One of the most important is the very slow movement of tectonic plates, which is the origin of the formation of continents, mountains, oceans and marine currents. The global climate has undergone significant changes through the entire history of the evolution of the planet. A relatively recent example of natural global changes are the glacial–interglacial cycles of the Pleistocene, with an average period close to 100,000 years. Their origin is related to small cyclical changes in the eccentricity of the Earth's orbit and the obliquity of the axis of rotation and also its movement of precession.

In the last millennia, and especially in the 20th and 21st Centuries, some human activities have generated environmental changes that have a global character due to their nature, their intensity and the size of their geographical distribution. In general, these anthropogenic global changes occur in periods of time much shorter than those of natural global changes. Because of these different timescales, it is often difficult to separate quantitatively the one from the other. Here, we will be interested above all in global anthropogenic change, which will simply be called global change, except when there is a question of distinguishing the natural effects from the anthropogenic effects. Human societies, through multiple activities carried out now and in the past, interfere with the Earth system in such an intense way and to such a temporal and spatial extent, that they threaten its various subsystems and the biotic and abiotic processes on which the sustainability of the societies themselves depend [STE 04].

The interference of human activities on the environment is, of course, felt at a local level (urbanization, pollution of the air and water, changes in land use, etc.), but it also has global consequences, in particular on the global climate and on non-climatic aspects, such as water and soil resources, biodiversity and natural resources.

In global change, it is useful to distinguish between systemic global changes, which manifest themselves directly on the Earth system scale, and cumulative global changes that occur at a local or regional scale, but which have a global expression because they arise almost all over the planet with such an intensity that they become a global problem. Global climate change caused by anthropogenic greenhouse gas emissions is a very important example of a systemic global change. However, there are others, such as the reduction in the concentration of stratospheric ozone, resulting mainly from the use of chlorofluorocarbons or CFCs, and the variations in albedo due to changes in land use and desertification. The main examples of cumulative global change are: water stresses caused by an overexploitation of water resources, the degradation and destruction of a large variety of ecosystems, biodiversity loss, soil retrogression and degradation, desertification, the disruption of the nitrogen and phosphorus cycles, pollution of the air, ocean, surface and ground waters, the increase in the concentration of aerosols of anthropogenic origin and, more generally, the increasing scarcity of renewable and nonrenewable natural resources.

4.2. Coastal zones and global systemic and cumulative changes

Among the systemic global changes, the only one that acts strongly on the coastal zones is anthropogenic climatic change. This change leads to increases in the global average sea level and in the sea temperature, to a reduction in the extent of sea ice and to changes in the acidity, salinity, wave climate and oceanic circulation. We will have the opportunity to analyze these effects in the following sections. According to the definition of the United Nations Framework Convention on Climate Change, anthropogenic climate change or simply climate change is the change in climate which is attributed directly or indirectly to human activity that alters the composition of the global atmosphere and which is in addition to natural climate variability over comparable periods of time. There are essentially two types of response to anthropogenic climate change: mitigation and adaptation. Mitigation measures are human interventions that seek to reduce the sources or enhance the sinks of greenhouse gases in the atmosphere; adaptation is the process of adjustment to actual or expected climate and its effects. In human systems, adaptation seeks to moderate harm or exploit beneficial opportunities. In natural systems, human intervention may facilitate adjustment to the expected climate and its effects. A system's vulnerability to climate change depends on the character, magnitude and rate of the change, the variations to which it is exposed, its capacity for adaptation and the present or future impacts on the system. These impacts depend in their turn on the sensitivity of the system. For human systems, the capacity for adaptation rests on several factors that belong to them:

social cohesion, economic resources and their diversity, political organization and living conditions. The concept of resilience, which means the capacity of an organism or a system to adapt to a changing environment and to maintain its essential function, identity and structure, is also often used instead of adaptive capacity although they are different concepts. We also speak of risk when faced with an uncertain hazard (a dangerous phenomenon or process) for a system or geographical zone where there are stakes that can be human, economic or environmental. Risk is often represented as a probability of occurrence of hazardous events or trends multiplied by the consequences if these events occur.

Regarding the effect of cumulative global changes on coastal zones, the question is more complex. Due to the fact that coastal zones are very dynamic regions where several drivers interact originating a large variety of different effects, attributing an origin to the impacts observed is difficult. The main driver that determines the dynamic of the anthropogenic global changes and their impacts on the coastal zones is socio-economic development and population growth. According to estimates for the year 2000 [MCG 07], the low elevation coastal zones, which only constitute 2% of the Earth's surface above sea level, shelter 10% of the world's population, corresponding to approximately 610 million people and 13% of the global urban population (about 360 million people). On a global level, about 65% of the world's cities with more than 5 million inhabitants, the majority of small island states and deltas with a high population density are situated in low elevation coastal zones.

In coastal towns, the risk of losses due to flooding, associated in some regions with subsidence, is currently increasing significantly due to increases in population and exposed assets, to urbanization and to the construction of infrastructure. According to Hallegatte *et al.* [HAL 13], the annual losses due to floods in the 136 most populous coastal cities were, in 2005, in the order of 6 billion dollars. Without new, efficient adaptation measures in these cities, the annual cumulative losses would reach more than 1,000 billion dollars in 2050. Even with adaptation measures, assuming a constant flooding probability over time, it can be estimated that the global sea level rise and the sinking of the seashore in some regions, will lead, by 2050, to annual losses of 60 to 63 billion dollars. Asia is the continent where the greatest number of people and the highest value of infrastructures and coastal assets are exposed to the flood risk. In the future, as a result of demographic growth and socio-economic development, the highest risks for coastal zones lie in Asia and Africa [HAN 11].

The socio-economic and climate scenarios indicate that in 2070, the countries most vulnerable to the risk of flooding will be: China, India, Bangladesh, Vietnam, the United States, Japan, Thailand, Myanmar, Egypt and Nigeria [HAN 11]. The majority of these countries are developing countries.

Among the most populous coastal cities where the potential losses due to floods will be the highest in 2050, can be found: Guangzhou, Mumbai, Kolkata, Guayaquil, Shenzhen, Miami, Tianjin, New York-Newark, Ho Chi Minh City and New Orleans [HAL 13]. The future situation is particularly worrying in the developing countries that combine a strong demographic growth and a weak capacity to adapt. In these countries, the ecosystems that protect the coast, such as wetlands, mangroves and dunes, are destroyed or their functioning is seriously affected by urbanization and development. It should also be noted that the migration of populations toward the coast is higher in the regions vulnerable to tropical cyclones and to the risk of flooding. It is important to recognize too, that these populations' current capacity for adaptation can improve rapidly with socio-economic development. The people who live in the shanty towns situated in the low elevation coastal zones, above all in the big cities of South and South-East Asia, are particularly vulnerable to the impacts of extreme meteorological events and to rising sea levels.

4.3. The impact of climate change on coastal zones

4.3.1. *Socio-economic and climate scenarios*

The preceding arguments show that there are strong connections and relationships of direct and indirect cause and effect between the socio-economic development of coastal zones and the vulnerability of these zones to climate change. As far as the long-term future is concerned, that is to say in the next 50 years and up to the horizon of 2100, there is the question of knowing toward what kind of climate the Earth is evolving. First, it is necessary to turn to the socio-economic scenarios that have been developed by the Intergovernmental Panel on Climate Change (IPCC) and called SRES scenarios [SRE 00]. The temporal trajectories for atmospheric concentrations of greenhouse gas resulting from a given SRES scenario are imposed as external forcing on the atmosphere–ocean coupled general circulation models that allow projections of the future climate to be made. Following the AR4 report – the IPCC's fourth evaluation report – scenarios for the concentration trajectories, called RCP for Representative Concentration Pathways [MOS 10], are used; these serve as forcing conditions for the socio-economic and climate scenarios. Each RCP scenario corresponds to a given total radiative forcing value for greenhouse gas emissions in 2100. Thus, the scenario RCP8.5 corresponds to a radiative forcing of 8.5 W/m^2 at the end of the 21st Century and represents the scenario where the volume of the emissions is higher during this century. Scenario RCP2.6 represents the development of measures for mitigation that reach the objective of not exceeding an increase in global average temperature higher than 2°C, compared to the pre-industrial era. The RCP scenarios produce data essential for climate scenarios, but for the study of vulnerability, impacts and adaptation, socio-economic scenarios are also needed. The studies carried out on the possible trajectories for socio-economic development up to the end of the century [MOS 10] have enabled five scenarios, called SSP for Shared Socio-economic Pathways, to be defined.

Scenario SSP1 corresponds to a situation where the global population is lowest and where the global GDP is highest; it is called "sustainable world" since it reflects a world that succeeded in making a transition toward sustainable development, addresses environmental questions, encourages innovation and technological development and avoids intensive consumption of natural resources, in particular fossil fuels. SSP3, called "fragmented world", reflects a world where the population is increasing rapidly, distributed between very poor regions with low consumption of natural resources and rich regions that depend heavily on fossil fuels. The inequalities are less accentuated in scenario SSP4, called "inequality". Scenario SSP2, called "middle of the road", prolongs the tendencies of the last decades and corresponds to the intermediary assumptions. Finally, scenario SSP5 corresponds to an increased consumption of fossil fuels and a weak awareness of environmental questions. Figure 4.1 presents different levels of engagement on mitigation and adaptation associated with the SSP scenarios. At the extremes are scenarios SSP1 and SSP3 with the greatest and smallest levels of engagement on mitigation and adaptation to climate change.

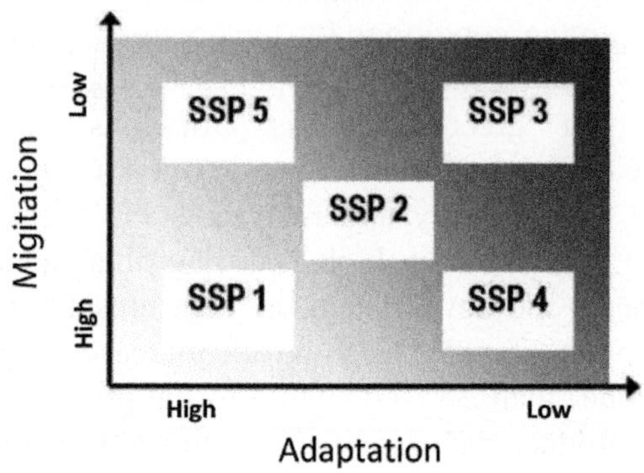

Figure 4.1. *Different levels of engagement associated with the SSP scenarios*

4.3.2. *Impacts of climate change on coastal zones on different time scales*

The changes resulting from the increase in concentrations of greenhouse gases in the atmosphere, their effects on different time scales on the different subsystems of the climate system and their impacts on the coastal zones will now be addressed. First, we will analyze the climate factors that have short-term impacts on the seashore. For example, the passage of a storm through the coast with a duration, normally, of less than a week, leads to a temporary elevation of the sea level that can cause temporary flooding or marine submersion in low elevation coastal areas and erosion of the seashore. Flooding is particularly serious if during the storm, the sea level reaches extremely high levels resulting from the coincidence of the storm surge and the high tide, especially at spring tides. Storm surges are associated with tropical cyclones and also with strong extratropical storms, especially those that have a rapid development and displacement. They trigger a rapid increase in sea level, due to the accumulation of water in downwind regions and the elevating effect triggered by the lower atmospheric pressure at the center of the depression (Figure 4.2). The accumulation of water near the coast can, again, be accentuated by Ekman pumping, by the shape of the seabed and by the funneling effects produced by some coastlines.

Figure 4.2. *Temporary elevation of sea level during a storm due to the coincidence of the storm surge and the high tide*

What interests us particularly here is the fact that the frequency of extreme local sea levels is increasing with the increases in local mean sea level (LMSL). This level is defined as the average sea level at a specified locality in the seashore, measured in relation to a reference point on land and over a sufficiently long period of time (a month, a year), so that the average value will be independent of fluctuations caused by waves and tides. The variations of the LMSL result from the combination of vertical movements of the seashore, either of elevation (of tectonic origin or caused by the postglacial rebound) or subsidence, with the increase in the global mean sea level (GMSL), triggered by climate change. The change in GMSL, or eustatic change, is normally measured in relation to the center of the Earth and results from a change in ocean water volume or in the volume of the oceanic basins. Since the Industrial Revolution, the GMSL has increased by about 20 cm. According to the IPCC AR5 report [IPC 13], during the 20th Century, the mean annual increase of the GMSL has been 1.7 mm, but since 1993 it has been between 2.8 and 3.6 mm. The same report considers that probably in the period 2081–2100, the GMSL will be 0.26 to 0.98 m higher than in 1986–2005.

On longer timescales, changes in the wave regime, resulting from climate change, cause erosion and accretion along the coast, which are variable at regional and local levels. Furthermore, the increase in the sea surface temperature changes the distribution of species in the medium and long term, leading to changes in the biodiversity of coastal ecosystems. The increasing CO_2 atmospheric concentration leads to an increase in the amount of CO_2 dissolved in seawater, where it forms carbonic acid. This phenomenon, which is called the acidification of the ocean, has the effect of decreasing the quantity of carbonate, and has harmful consequences for marine biodiversity, such as depressing metabolic rate and immune response in some organisms, and causing coral bleaching. The combined effect of increasing sea surface temperature and decreasing pH of the Earth's oceans is very likely to have harmful consequences on the physiological functions of numerous marine organisms.

Currently, the increase in GMSL is changing the frequency of the most extreme sea levels during tropical and extratropical storms. There are other potential impacts however. The rise in GMSL carries significant risks for natural and human coastal systems at a regional and local level, specifically a higher risk of flooding, erosion and salinization of water resources and coastal agriculture soils. In the long term, up to the end of the century, the increase in GMSL will very probably become a major problem in low-lying coastal zones. The situation will be particularly difficult for countries with a weak capacity to adapt such as for example, small island developing states. We can expect migratory movements from developing countries where large populations are concentrated in coastal zones vulnerable to flooding. The problem will not stop at the end of the century. It is very probable that the rise in GMSL will continue over several centuries because of the thermal inertia of the ocean and the persistence of the melting of the polar ice in Greenland and Antarctica. It is estimated that the current atmospheric concentration of greenhouse gases leads to an inevitable rise in the GMSL by 2.6 m in the next 2,000 years [LEV 13]. To respond to these challenges, it will be necessary to develop adaptation strategies embracing several centuries.

4.3.3. *The rise of the global mean sea level – observations and projections*

On geological timescales, the GMSL has been affected by the formation of polar icecaps during former glaciations (the last began during the Carboniferous period and continued into the Permian, 360 to 260 million years ago), by changes in the shape of oceanic basins and by change in the configuration of the continents. During the quaternary glaciation, the GMSL varied by about 100 m in the glacial–interglacial cycles of the Pleistocene. These changes are principally due to the periodic increase and decrease in the polar caps' ice mass, associated with changes in the global mean temperature of the atmosphere. Since the peak of the last glacial period around 18,000 years ago, the GMSL has risen by approximately 120 m. However, during the last interglacial period around 129 to 116 thousand years ago, the global mean temperature of the atmosphere at the surface being around 2°C greater than the corresponding temperature in the pre-industrial era (before the middle of the 18th Century), the GMSL was very probably more than 5 m higher compared to the current value [IPC 13]. Recent reconstructions of the evolution of the GMSL during the last 2,100 years [KEM 11], represented in Figure 4.3, indicate a stable period between 100 B.C and 950 A.D., followed by a rise of around 0.2 m up to the end of the 20th Century. After this period, the GMSL has risen by an average of 2.1 mm per year, which represents the highest rate of increase in the last 2,000 years.

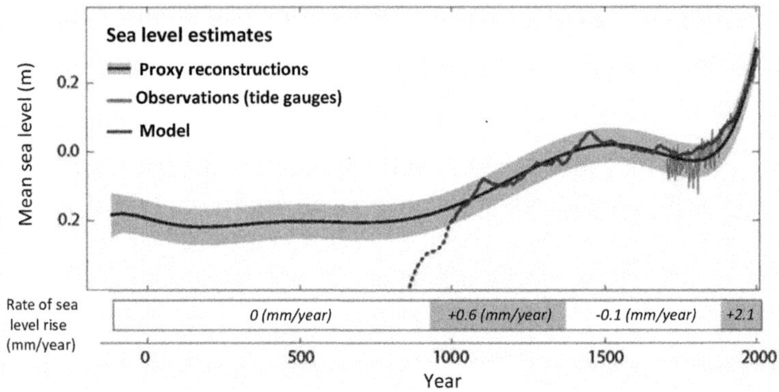

Figure 4.3. *Evolution of GMSL over the last 2,100 years (adapted from [KEM 11]) (see color section)*

The variation in GMSL has been measured since the 18th Century by tide gauges and since 1992 by satellites using radar altimeters, in particular by the satellites TOPEX/Poseidon and Jason 1 and 2. According to the IPCC AR5 report [IPC 13], during the 20th Century the mean annual increase of the GMSL was 1.7 mm, but since 1993 it has been situated between 2.8 and 3.6 mm. We are therefore observing an accelerated rise of the GMSL that has recently been quantified [CAL 13].

The increase in mean sea level is not spatially uniform. For example, the rate of increase in the mean sea level in the western Pacific equatorial region is around three times greater than the rate of increase in GMSL. However, on the eastern Pacific equatorial region, the rate of increase is less than the global rate, as Figure 4.4 shows for the period 1993–2012. These spatial variations of the local mean sea level, relative to the center of the Earth, occur on small timescales – from a day to several years – and are associated with storms and climate variability, especially with the El Niño southern oscillation and the Pacific decadal oscillation. The climate variability induces changes in winds at the ocean surface, changes in oceanic currents, in sea surface temperature and salinity. All these changes lead to a variation in the local mean sea level. On this short-term spatial variability is superimposed the long-term variation resulting from anthropogenic climate change and acting on the same parameters, the winds, oceanic currents, temperature and salinity.

When considering the impacts of the mean sea level rise on a specific location on the coast, it is necessary to know the rate of change of the local mean sea level and the rate of change of the vertical movements of the land at the same location. Figure 4.4 shows, for example, that in Stockholm the mean sea level is decreasing, whereas in Manila it has been rising for 40 years at a faster rate than the rate of rise in GMSL. In conclusion the rate of variation in LMSL is strongly dependent on the region under consideration. However, it is very likely that by the end of the century, about 95% of the ocean's surface will show an increase in its mean level. The ocean regions where a decrease is not likely will be those close to glaciers, polar ice caps and ice sheets [IPC 13].

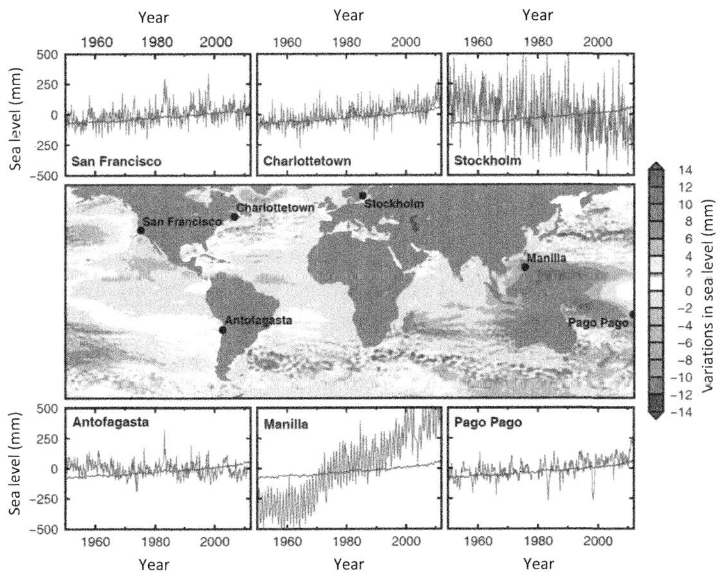

Figure 4.4. *Rates of change of mean sea level in relation to the center of the Earth for the period 1993–2012. The figure also shows the variations (in grey) of the LMSL in six coastal towns, detected by tide gauges, for the period 1950–2012. The lines in red represent an estimate of the variation in GMSL in the same period (adapted from [IPC 13]) (see color section)*

Projections

To make projections on the future rise in GMSL, it is necessary to fully understand the mechanisms that are at work. Two main processes, due to climate change, cause the rise in GMSL: the thermal expansion of the upper layers of the ocean and the melting of land ice on glaciers, polar ice caps and ice sheets. A third process results from human intervention in the storing of water on land, associated mainly with the construction of dams that retain water and the exploitation of aquifers. We have here a good example of the combined effect of a global systemic change (climate change) and a global cumulative change (the exploitation of surface and ground waters on a global scale). The retention of water in the roughly 30,000 dams, constructed globally during the 20th Century, caused a small decrease in GMSL between 1940 and 2007, with an estimated total value of 23 mm [CHA 08]. Following increasing demand for water, the exploitation of aquifers and the return of the used water to the hydrological network at present cause a small increase in sea level. Currently, the net mean annual variation in GMSL due to these intensive uses of freshwater is positive and estimated at 0.38 mm, for the period 1993–2010 [IPC 13].

The two contributions of anthropogenic climate change to GMSL rise will now be analyzed further. The rise due to thermal expansion is approximately proportional to the increase in thermal energy in the ocean, with a proportionality constant of 0.11 m per 10^{24} J [KUH 12]. It is important to further analyze the origin of this increase in energy. The increasing concentration of greenhouse gases in the troposphere, or lower atmosphere, due to certain human activities, means that the radiative forcing at the top of the troposphere is increasing. This implies disequilibrium in the climate system that leads to an increase in its thermal energy. It is estimated that, during the last decades, 93% of that energy excess has been transferred to the ocean [IPC 13] because its thermal capacity is much larger than the other subsystems of the climate system. The remainder of the energy excess warms the atmosphere and the continents and causes the melting of ice in continental glaciers, ice sheets, ice caps and ice shelves. It can, then, be concluded that in the absence of the ocean or with a smaller ocean, the increase in the global mean temperature of the atmosphere at the surface would have been much larger than the increase observed. The ocean therefore mitigates, quite considerably, the effect of anthropogenic greenhouse gas emissions on the temperature of the Earth's atmosphere.

Another very important aspect is that the oceans and the cryosphere are the main sources of physical inertia in the climate system, for timescales in the order of millennia. If we manage to reduce the anthropogenic greenhouse gas emissions considerably, the positive radiative forcing on the climate system ceases to increase and stabilizes. Yet it will take decades, indeed centuries before the mean temperature of the atmosphere close to the surface reaches an equilibrium temperature. The ocean's stabilization period is even longer due to its relatively high thermal capacity, and the slow pace of vertical thermal exchanges. The heat propagation from the surface to the oceanic depths occurs very slowly, taking centuries to millennia. Consequently, the increase in the volume of the ocean due to its thermal expansion is a very slow process that will last for centuries and possibly millennia. The response of the cryosphere to the positive radiative forcing is slower and will probably last even longer.

However, not all the impacts on the cryosphere caused by climate change lead to an increase in GMSL. The melting of sea ice, especially in the Arctic, does not contribute to it, except for a slight increase linked to the difference in density in the water from the melted sea ice and the sea water. Two types of floating ice can be distinguished in the polar regions: the vast ice shelves on the coasts of Antarctica, that form where the glacier reaches the sea, and the sea ice that covers a large part of the Arctic.

Model-based estimates indicate that, in the period 1993–2010, the thermal expansion of the upper layers of the ocean contributed an average 1.1 mm per year to the rise in GMSL, the melting of glaciers, 0.86 mm, the melting of the Greenland ice sheets, 0.33 mm and the melting of the Antarctica ice sheet, 0.27 mm. When 0.38 mm of GMSL rise, resulting from the exploitation of aquifers is added to the equation, a total of 2.94 mm is obtained. This result is in fairly good agreement with the direct observations from tide gauges and satellites equipped with altimeters, which since 1992 have measured an average annual rate of increase in GMSL between 2.8 mm and 3.6 mm [IPC 13]. The difference between the values calculated and observed probably results from the fact that the contribution from the melting of the polar ice sheets is greater than the models' estimate. Nevertheless, the agreement obtained constitutes a validation of the observations and of the models and finally of our interpretation and knowledge of the origins of GMSL rise. It can also be concluded that the observed evolution of the GMSL follows the projections that have been made successively by the four previous reports of the IPCC. This agreement, represented in Figure 4.5, also constitutes a validation of the interpretation of the GMSL rise over the preceding two decades.

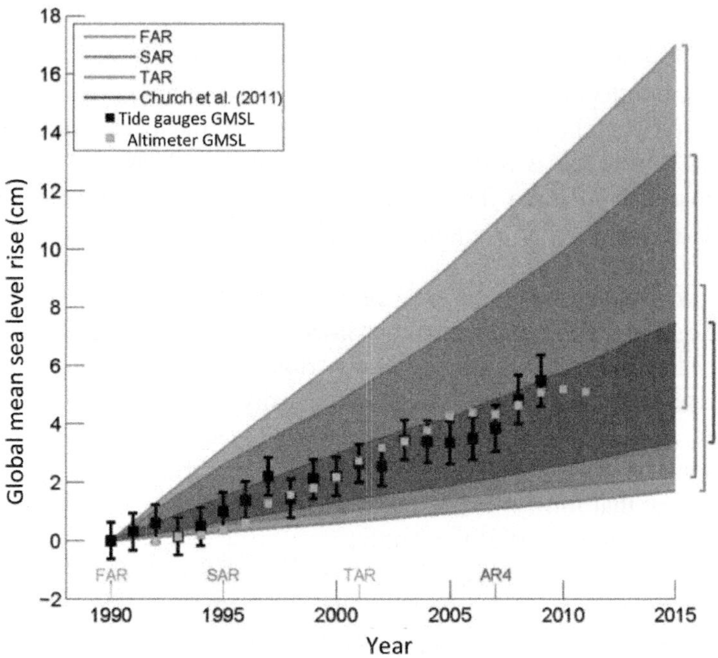

Figure 4.5. *Variation in GMSL from 1990 to 2011, evaluated using data from tide gauges (error bar in black) [CHU 11] and satellites TOPEX and Jason (blue squares) [NER 10]. The colored parts represent the margins in the projections for the GMSL from IPCC reports 1° (FAR), 2° (SAR), 3° (TAR) and 4° (AR4), corresponding to the different scenarios for emissions of greenhouse gas (adapted from [IPC 13]) (see color section)*

In regards to the future, there are two main methods for making GMSL projections. Models used are based either on the modeling of geophysical processes in the various systems (upper layers of the ocean, glaciers, ice sheets and storing of land water resources), or semi-empirical models. The latter project the future behavior of GMSL by using statistical relationships between observations on past behavior, and the mean global temperature of the atmosphere at the surface [GRI 10, RAH 07] or the radiative forcing [JEV 10].

The models based on geophysical processes project for the period of 2081–2100, compared to 1986–2005, an average increase in GMSL between 0.26 and 0.55 m for scenario RCP2.6, between 0.32 and 0.63 m for RCP4.5, between 0.33 and 0.63 m for RCP6.0 and between 0.45 and 0.82 m for RCP8.5 [IPC 13]. Figure 4.6 shows the evolution of the GMSL rise projected by the four RCP scenarios.

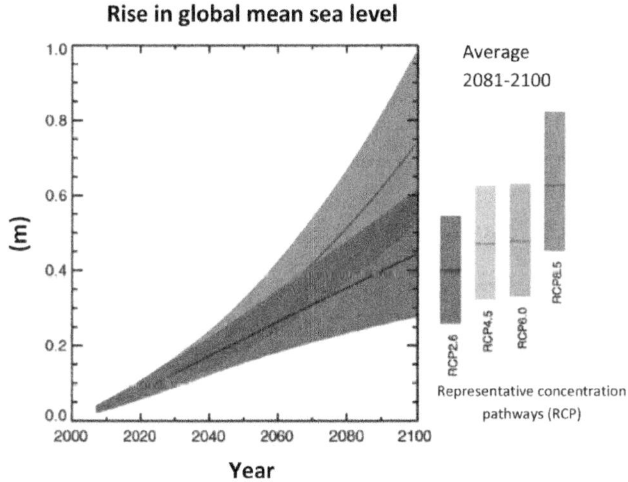

Figure 4.6. *Projections for the rise in GMSL up to the end of the century, resulting from different RCP scenarios [IPC 13]. The margins in blue and red correspond to the uncertainties of the extreme scenarios RCP2.6 and RCP8.5 (see color section)*

For the RCP8.5 scenario, which corresponds to the larger volumes of global emissions of greenhouse gas, the models project a GMSL rise in 2100 between 0.45 and 0.98 m and a mean annual rate of GMSL rise, for the period 2081–2100, of 8–16 mm, which is more than double the current mean rate. If we do not succeed in reducing current global annual emissions by 60–80% before 2050, at the latest, it is very probable that the GMSL rise will reach more than 0.5 m at the end of the century. This projection, in conjunction with a very probable intensification of extreme meteorological events, in particular tropical cyclones and extratropical storms, will lead to extreme sea levels that will be much more frequent and higher. This tendency is potentially very destructive for low elevation coastal zones, especially for those where tropical cyclones are frequent.

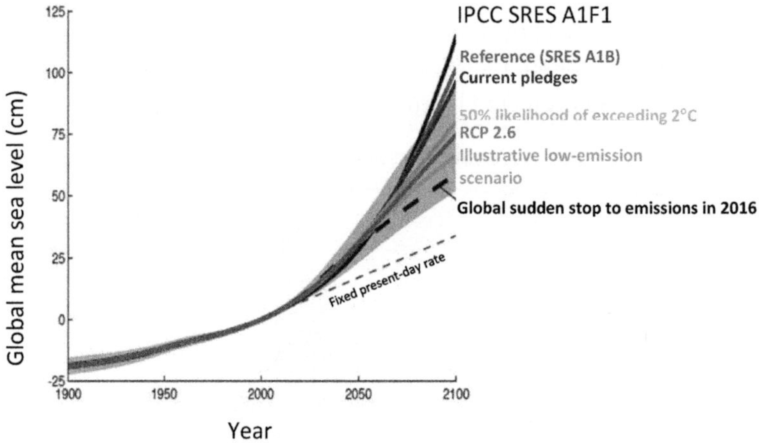

Figure 4.7. *Evolution of GMSL projected by the semi-empirical models with different scenarios for emissions described in the text. According to these models, the current commitments for mitigation are insufficient to avoid a rise in GMSL lower than 50 cm at the end of the century (see color section)*

The projections from semi-empirical models lead in general to GMSL rises higher than those from modeling of geophysical processes. Often, such projections are larger than 1 m at the end of the century [VER 09]. In Figure 4.7, the evolution of GMSL is represented, as projected by semi-empirical models, with different emission scenarios. The broken line represents the rise if the current annual rate is maintained; the broken line in bold corresponds to a scenario where greenhouse gas emissions cease abruptly in 2016. Note that the rise in GMSL is maintained by the thermal inertia in the ocean. Even if it is imagined that all greenhouse gas emissions cease abruptly in 2016, the GMSL would continue to rise for centuries or millennia. The green line refers to a scenario where the global emissions are reduced by 80% before 2050. Current engagements correspond to reductions in emissions agreed voluntarily by the countries that have established quantified mitigation. The four other lines correspond to different SRES [SRE 00] scenarios, to scenario RCP2.6, and to a scenario where the probability of not exceeding an increase of 2°C in the mean global temperature is 50%.

There are still considerable uncertainties over the effect of global warming on the behavior of polar caps. Recent observations show that changes in the dynamic of the polar ice have often been under-estimated. From the beginning of the 1990s, the loss of ice mass in the Arctic and the western Antarctic ice sheets has accelerated considerably. In the period 1992–2011, the mean annual rate of ice mass loss has been 142 gigatonnes in Greenland and 71 gigatonnes in the Antarctic. [SHE 12]. The models currently available indicate that beyond a certain threshold of increase in the mean global temperature of the atmosphere, global warming will lead to an irreversible and almost complete disappearance of the Greenland ice sheet. Calculations show that this threshold is situated between 2 and 4°C compared to the pre-industrial temperature [RID 10]. The process of ice melting in Greenland will last more than a millennium leading to a rise in GMSL close to 6 m.

The Antarctic ice sheet is more stable, particularly in its eastern part. It is estimated that the melting of the whole Antarctic ice sheet would contribute to a rise in sea level of more than 60 m. The western part of the Antarctic is less stable and, if it melted, would lead to an increase in GMSL of nearly 4 m. It is important to note that a large proportion of the western Antarctic ice sheet is situated below mean sea level, which renders it particularly unstable in a case of continuing global warming. Although the uncertainty over the future behavior of this part of the Antarctic is considerable, its melting is likely to begin before the end of the century [IPC 13].

4.3.4. *Other impacts of climate factors on coastal zones*

4.3.4.1. *Tropical cyclones and extratropical storms*

When tropical cyclones and strong extratropical cyclones strike the seashore, the storm surges produce a temporary rise in sea level that, coupled with the high waves, can cause devastating floods in coastal zones. Tropical cyclones are convective mesoscale systems that form in the oceanic intertropical regions and generate heavy rains, strong winds that can reach more than 250 km/h and high waves. It is difficult to reconstruct, reliably, the evolution of their frequency and intensity during the 20th Century because the methods, regularity and quality of the observations have changed a great deal. Nevertheless, it is very probable that the intensity of the strongest cyclones (those where the maximum sustained wind speed is higher) have been increasing, especially in the North Atlantic [WEB 05]. It has also been observed that tropical cyclones become more frequent during years that are warmer globally and that the strongest are more sensitive to the increase in temperature [GRI 12]. It is estimated that global warming will double, by the end of the century, the number of cyclones with the intensity of Katrina, one of the strongest cyclones in the North Atlantic, which occurred in August 2005 and accounted for 1,800 victims in the United States, where it caused around 100 billion dollars (in 2005) worth of damage [GRI 13].

The projections based on high-resolution dynamic models indicate that anthropogenic climate change tends to decrease the mean global number of tropical cyclones by 6 to 34% [KNU 10]. The same models project an increase in the mean global intensity of tropical cyclones by 2 to 11%. This implies that in the future, the frequency of very intense cyclones is likely to be higher, which will aggravate their impact on coastal zones by increasing the flood risks associated with storm surges, high waves and rising GMSL. When considering the impacts of tropical cyclones in low elevation coastal zones it is necessary to distinguish between the climate drivers and the non-climate drivers. According to a recent United Nations report [GAR 11], the population exposed to the risks of tropical cyclones has increased by 192% in the period 1970–2010, while the global population has increased by 87%. Nevertheless, since 2000 the mortality due to tropical cyclones has decreased in all regions of the world for various reasons, in particular the implementation of warning systems. However, the risk of economic losses has increased in less developed countries, particularly in Africa and Asia, where projections based on socio-economic scenarios indicate that this risk will continue to increase, in part for reasons linked to the adoption of development models that do not decrease the vulnerabilities of the coastal zones. One example where these different risk factors were combined was the super typhoon Haiyan, of category 5 on the Saffir-Simpson scale, which generated a storm surge of about four meters height in the island of Leyte in the Philippines. This tropical cyclone ravaged the coasts of the islands in the central area of the Philippines in November 2013, caused more than 5,000 deaths and displaced tens of thousands of people, mainly in the coastal shanty towns (Figure 4.8).

Figure 4.8. *Destruction caused by typhoon Haiyan (November 2013) in the coastal shantytowns of the town of Tacloban in the Philippines (photo: The Weather Channel)*

In the middle latitudes, storms are associated with the general circulation of the atmosphere. Since the 1970s the tropical belt has widened, storm tracks and jet streams have shifted toward the poles which implies a migration of the trajectories of extratropical storms to higher latitudes. Yet, the observations and the climate scenarios have not been able to determine clear changes in the intensity and frequency of storms in the middle latitude regions [IPC 13].

4.3.4.2. *Winds, wave climate and sea currents*

Wave formation results from the action of winds on the ocean's surface. When the wind waves propagate over large distances, they form the swell. The dissipation of energy when the waves break on the seashore defines the coastline by transporting unconsolidated sediments and controls the phenomena of erosion, floods and submersions. Climate change, by modifying the wind regimes and the trajectories, intensity and frequency of storms, changes the wind wave climate and the swell, with very significant effects on the coastal zones. For example, a rotation of the mean direction of the waves can modify considerably the transport of sediments along the coast and aggravate the erosion [SAN 02]. A change in the wind regime on the coast can also change the intensity of upwelling, with significant effects on biological production and fishing stocks.

Considerable uncertainties remain regarding future projections for the mean and extremes of winds, waves and swell on the coast, especially at a regional and local level, obtained from climate models [IPC 13].

The changes in the general circulation of the atmosphere resulting from global warming can also change the intensity and direction of marine currents. For example, the thermohaline circulation in the North Atlantic that transports heat northward and contributes to maintain relatively mild temperatures in North-West Europe will very probably slow in the course of the 21st Century. There too, uncertainty remains over the evolution of this phenomenon and more generally over the impacts of climate change on global and regional oceanic circulation.

4.3.4.3. *Sea surface temperature*

It has already been underlined that currently the ocean absorbs more than 90% of the excess thermal energy resulting from the intensification of the greenhouse effect. This absorption leads to a warming of the ocean, especially of its upper layers above 700 m in depth. The average rate of warming is higher close to the surface where it has increased by 0.1°C per decade, in the first 75 m, over the period 1971–2010 and decreases in depth [IPC 13]. Below 700 m, a systematic warming of the ocean has already been observed up to 2,000 m in depth. The warming of the ocean is on average stronger near the coast.

Extreme values of the sea surface temperature, or "maritime heat waves", have been recorded on the coasts of West Australia with temperatures of 5°C above average [PEA 13]. This warming of the upper layers of the ocean currently pushes marine and coastal species nearer to the poles at an average estimated speed of 72 km per decade [POL 13]. This rate of migration is considerably larger than that of land species, which migrate toward the poles at an average speed of six kilometers per decade. It has also been shown that on average the seasonal cycle of marine life begins four days earlier and that the algal blooms occur six days earlier than in the past, whereas fish larvae appear eleven days early [POL 13].

The warming of the ocean also affects the coral reefs and increases the frequency of algal blooms that are, very often, intensified by the supplies of nutrients of anthropogenic origin; these blooms can upset the equilibrium of the food chain at a local level and can lead to the prohibition to harvest or sell some seafood species.

Finally, let us remember that the increase in the temperature of the upper layers of the ocean contributes to the water stratification and modifies the dynamics of the ocean currents and of the ice in the polar regions.

4.3.4.4. *Ocean acidification*

With the increase in the atmospheric concentration of CO_2, the flux of absorption into the ocean becomes larger than the inverse flux of emission into the atmosphere. The dissolution of some of the excess CO_2 into the ocean leads to its acidification [GAT 13]. The pH at the ocean surface, which had the value of 8.25, has decreased by around 0.1 since the industrial revolution and is currently 8.14, which corresponds to an increase in concentration of the ion H^+ by 26%.

Ocean acidification is a direct consequence of the CO_2 anthropogenic emissions and not of the climate change resulting from such emissions. The ocean functions as an enormous sink for CO_2 which, since the industrial revolution, has absorbed about 30% of the total anthropogenic emissions of this gas. If this sink did not exist, the concentration of CO_2 in the atmosphere would be much higher. We find here another example where the ocean plays the role of climate regulator. Contrary to other effects of climate change, ocean acidification is immediately reversible and could be directly controlled by limiting the CO_2 emissions into the atmosphere.

All RCP scenarios project a decrease in the ocean's surface pH for 2100, with values situated between 0.06 and 0.32; for the extreme scenarios RCP2.6 and RCP8.5, these decreases range from 0.06 to 0.07 and from 0.30 to 0.32, respectively.

Ocean acidification results mainly from a global systemic change but also from the presence of nitrogen and sulfur of anthropogenic origin in coastal waters, associated with cumulative global changes. Even if this last process has a weak impact on a global scale, of the order of a few percent in relation to the contribution of atmospheric CO_2 [DON 07], it is nevertheless responsible for significant temporal and spatial variations that have harmful effects on biodiversity and coastal and marine ecosystems.

4.3.4.5. *Ocean deoxygenation*

In the first 100–200 m of the ocean's water column, oxygen is generally abundant as a result of dissolution from the atmosphere. Moreover, photosynthesis in the phytoplankton produces large amounts of oxygen with seasonal variations. The oxygen dissolved in the ocean is used by organisms that consume organic matter, which comes directly or indirectly from phytoplankton. In the intermediary waters, the oxygen concentration decreases, but increases again beyond 1,000–2,000 m.

Today, several coastal zones, called hypoxic zones, have such low oxygen concentration that animal life becomes impossible, and as a result are also called "dead zones". The number of dead zones has practically doubled each decade since 1960 [DIA 08]. Ocean deoxygenation is mainly caused by the massive influx into the ocean of nitrate fertilizers, used in intensive agriculture, of untreated wastewater and indirectly by the use of fossil fuels. These processes enrich the coastal waters with nitrogen and phosphorus, nutrients that lead to the overproduction of algae and cyanobacteria the decomposition of which favors the growth of bacteria that consume the dissolved oxygen.

Anthropogenic climate change also contributes to ocean deoxygenation for two main reasons: first, because warmer water can hold less dissolved oxygen than colder water and second, because global warming, by contributing to the stratification of the upper layers of the ocean, makes it more difficult for oxygen to reach the deeper ocean. The deoxygenation of the ocean, which is stronger in the coastal waters than in the open sea, reduces the production of phytoplankton, at the base of the food chain, whereby it leads to negative impacts on fishing stocks.

In certain marine regions, the combination of ocean warming, ocean acidification, ocean deoxygenation, ocean pollution resulting from certain human activities and overfishing constitutes a very serious menace for the ocean and for the sustainability of the services rendered by the marine ecosystems.

4.3.4.6. *Impacts of precipitation regime changes*

Climate change scenarios project an increase in average precipitation over high latitudes, especially in the northern hemisphere, and over some of the regions of intertropical convergence, but predict a decrease over several regions in the middle latitudes, such as the Mediterranean Basin, the south of Europe, the south-east of the United States, the north of Mexico and the south of Africa. There is also a tendency toward an increase in the frequency of extreme precipitation, in particular droughts and heavy precipitation events followed by flooding. In mountains, the current tendency toward a decrease in the time period of snow coverage, a reduction in the volume of snow and in the volume of glaciers will be aggravated. The combination of these climate change impacts with the overexploitation of water resources in some regions of the world will change the flow regime of rivers and streams, in particular its seasonal behavior, the frequency and intensity of floods and drought events. There will be a progressive change in the quantity and quality of water reaching the ocean, although very variable at a local and regional level; these changes will have consequences on the transport of sediments and contaminants to the sea, on the salinity of coastal waters and on the estuarine and coastal ecosystems.

4.4. Impacts of cumulative global changes on coastal zones

In the preceding pages, several impacts of cumulative global changes on coastal zones have been addressed. One of the most important is the impact of the exploitation of surface and underground water resources. On a global level, the water retention in dams and the exploitation of aquifers contribute to changes in GMSL. It has also been noted that the exploitation of water resources changes the river regimes and influences the quantity and quality of the water that reaches the estuaries and the seashore. In developing countries, more than half of the used water flowing into the ocean is untreated; in Latin America and in the Caribbean, the percentage is close to 85%; in East Asia, it reaches 90%, and 80% in Central and Eastern Africa. In the South-East Pacific, it exceeds 80% [GAP 06].

The anthropogenic perturbation of the nitrogen and phosphorus cycles also has a strong impact on the coastal zones. It has been shown that the intensive use of fertilizers in agriculture and the discharge of urban waste in rivers, streams and lakes, leads to the eutrophization of estuaries, gulfs and bays downstream from drainage basins and also contributes, to the ocean deoxygenation. Aquaculture can also have negative impacts on the coastal environment, in particular if it leads to organic, chemical or genetic pollution.

In several regions of the world, the increasing organic and inorganic pollution of water resources and soils, originating from demographic growth and rapid industrial expansion without environmental protection, seriously threatens the ecological equilibrium of coastal zones. The presence of pollutants in coastal ecosystems can have repercussions on marine food systems and constitute a threat to the health of local and regional communities, affecting many activities, including tourism. Nevertheless, encouraging progress has been made to control and reduce the presence of persistent organic pollutants, radioactive substances, oil pollution and oil spills.

The construction of dams and other infrastructures on drainage basins and coastal plains, on a global scale, has reduced significantly the transport of sediments into coastal zones. This sediment deficit intensifies coastal erosion and threatens ecosystems, in particular mangroves. The large-scale extraction of sand, gravel or pebbles from rivers, estuaries and coastal zones for the building sector often contributes to coastal erosion.

4.5. Vulnerabilities of humans and natural coastal systems

A system's vulnerability to climate change depends on present and future impacts and on its capacity to adapt. When considering the vulnerability of human coastal systems, its three components, described in section 4.1.1, should be considered. The third component, namely the institutional, legislative, judicial and cultural activities, is essential to promote the climate change adaptive capacity that will be analyzed in section 4.6. The impacts on the first component – built environment, transport and industrial infrastructures and port facilities – result mainly from flooding, from land loss due to GMSL rise and from damage caused by extreme weather events, such as tropical cyclones and strong extratropical storms, heavy rains, extreme drought and heat waves. All these climate factors also have impacts on health, the availability of freshwater water and the food security of coastal populations.

The vulnerability of coastal anthroposystems to climate change is also increasing due to non-climate factors, such as, population growth, urbanization and high rates of subsidence caused by human activities in several low elevation coastal regions, in particular the deltas of the Nile, the Ganges and the Mekong. Projections based on socio-economic scenarios indicate that the global population exposed to a level of coastal flooding with an average return period of a hundred years will increase from 271 million in 2010 to 345 million in 2050 [JON 12].

Turning to natural coastal systems, it is important to note that their functioning is governed by complex physical and biogeochemical processes that are still insufficiently known. The overexploitation of resources, in particular water, sand, gravel and pebbles, the disruption of the nitrogen and phosphorus cycles, the pollution of water and soils and the degradation and destruction of habitats are at the root of the main anthropogenic impacts on coastal zones.

Coastal zones are dynamic environments, often rich in biodiversity, that are particularly vulnerable to climate change and to global cumulative changes. They are subject to many pressures, from the exploitation of mineral and living resources to the recreational, residential and tourism activities that they support. Beaches, dunes, barrier islands, lagoons, wetlands and deltas are the most threatened and should be the subject of adaptation strategies for climate change and sustainable development.

4.6. The adaptation of coastal zones to environmental changes

4.6.1. *Options and strategies for adaptation*

Coastal zones' adaptation to climate change is considered here specifically, but the adjustment of natural and human coastal systems to cumulative global changes and to the interactions and synergies between the two adjustment processes will also be addressed.

The concept of adaptation has already been defined in section 4.2; adaptation can be spontaneous or planned. The first case is adaptation in response to the experienced climate and its effects, without planning explicitly or consciously focused on addressing climate change. The process of spontaneous adaptation, also known as reactive adaptation, is not preceded by an analysis of climate scenarios, vulnerabilities and present and future risks. Planned adaptation, also called anticipatory adaptation, is on the contrary a strategic process based on knowledge of future climate scenarios and their impacts on natural and human systems. Its objective is the integration of the potential climate change impacts in management policies on a local, regional, national and international level, in order to moderate harm and exploit beneficial opportunities. Maladaptive actions or maladaptation are actions that may lead to increased risk of adverse climate-related outcomes, increased vulnerability to climate change, or diminished welfare, now or in the future.

Adaptation is a dynamic and continuous process that seeks to respond to the challenges posed at a local level by a climate that is in the process of changing. It should be understood as a policy of transition in the long-term, formulated in a cyclical fashion by the elaboration of studies on impacts and vulnerabilities, based on regionalized climate scenarios, on adaptation strategies, and finally on monitoring and evaluation programs. The repetition of cycles allows for the adaptation processes to be renewed and implemented with the incorporation of new scientific knowledge, new methodologies and new technology.

One of the major difficulties for adaptation is accounting for uncertainty in decision-making. There are uncertainties in the socio-economic scenarios, in the general circulation models used to generate the future climate scenarios and in the way these scenarios are regionalized to smaller spatial scales. There are also uncertainties in the models that estimate the impacts and vulnerabilities of natural and human systems and finally in the identification and selection of adaptation measures. Yet, it is necessary to recognize that uncertainty is an inevitable element in all decision-making processes. Rather than constituting a barrier to taking decisions, the knowledge, evaluation and integration of uncertainties increases the robustness of the decision-making process and reduces the risk of error.

The adaptation of coastal zones to climate change is a very general concept that reflects a vast range of activities, such as the identification and analysis of adaptation options, the implementation of the adaptation measures and the monitoring and evaluation of the adaptation process. The main adaptation options can be organized into three main categories: institutional and social options, technological options supported by engineering, and ecosystem-based adaptation options. The institutional options include the financial instruments, insurance, legislation, regulation, plans, programs and the participation of stakeholders in decision-making. The complexity of coastal zones' adaptation to climate change implies that the selection of options is strongly influenced by the educational and cultural level as well as the social cohesion of coastal populations. The outreach, sharing of information and awareness of the coastal zones vulnerabilities to climate change among local communities and stakeholders, and also the acquisition of in-depth knowledge on the different adaptation options, are an essential requirement for developing a successful adaptation process at the local level [SCH 13]. The ecosystem-based adaptation uses the potential contribution of biodiversity and ecosystem services (for example, sand dune ecosystems, wetlands, coral reefs and mangroves) to develop an adaptation strategy. By acting in concert with nature and by choosing ecological options, we have more chance of developing an adaptation in harmony with the complex coastal zones dynamics.

The classification of coastal adaptation strategies consisting of the trilogy of retreat, accommodation and protection is now widely used (Figure 4.9).

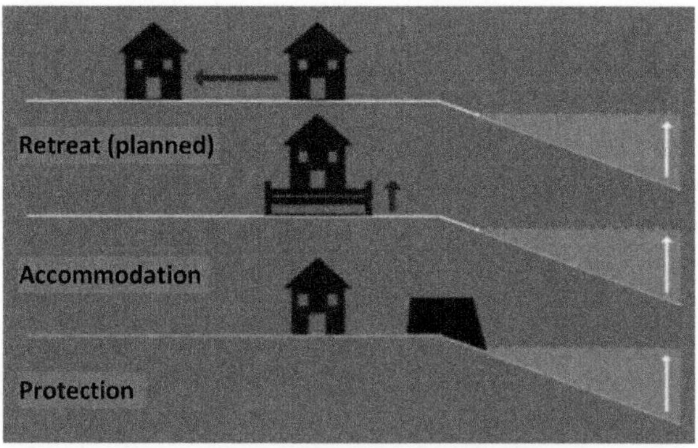

Figure 4.9. *The three adaptation strategies to rising GMSL in coastal zones*

Protection consists of advancing or holding the coastline using various options, such as beach and dune nourishment, ecosystem restoration and the construction of hard coastal defense structures. These can be seawalls, sea dykes, groynes and storm surge barriers. While seawalls aim to hold the coastline and to avoid flooding or submersion of low-lying land, transverse structures retain part of the sediments moving parallel to the coast. The hard engineering options are in general relatively costly, complex and have a limited lifespan. Planning of such options should take account of the uncertainties associated with the socio-economic and climate change scenarios and the side effects that they can have on the medium- and long-term local and regional coastal zone dynamics.

Accommodation privileges the change in human coastal activities and flexible adaptation of the built environment and infrastructures, such as the adaptation of buildings to make them more resistant to the effects of GMSL rise, the lifting of bridges and other transport infrastructure, the construction of flood shelters, flood proofing and the implementation of flood warning systems for extreme weather events leading to extreme values of sea level. This strategy also consists of land-use planning adaptations to reduce the flood risk and the development of new insurance programs. It also includes flood-resistant agriculture using plants that tolerate high levels of salinity. Accommodation favors flexibility and evaluation of the adaptation measures, while taking account both of uncertainties and additional future information on the evolution of vulnerabilities.

Retreat is a strategy for managing shoreline setbacks and realignment and the inland migration of coastal ecosystems to make them less vulnerable to erosion and to the rise in GMSL. For human systems and more specifically for the human occupation of the coastal zone, retreat is an extreme strategy that applies only when all the other options become impossible.

Nevertheless, it is very probable that there will be climate refugees coming from very vulnerable coastal zones and that their number will increase significantly in the future [UNE 11]. A current example is taking place on the Carteret Atoll, a group of six small islands belonging to Papua-New-Guinea, with a maximum elevation of 1.5 m above mean sea level and with a population of about 1,700 people. The floods associated with extreme weather events, particularly tropical cyclones, cause the salinization of soils and water resources, making agriculture and the population's survival very problematic. Some of the inhabitants of the Carteret Atoll have migrated to other less vulnerable nearby islands since 1980.

4.6.2. *Systems for analyzing institutional and government decisions*

When choosing from diverse options for the adaptation of coastal zones to climate change, it is generally necessary to consider long-term investment (more than 30 years) options, such as changes in land planning, the construction of defenses against flooding and submersion and construction of housing, leisure, tourism and transport infrastructures. This implies that classical cost-benefit analysis methods cannot be used because it is not possible to attribute probabilities to the different socio-economic scenarios that determine the scenarios for greenhouse gas emissions [LEM 01]. These are the scenarios that allow us to project the impacts and the vulnerabilities of long-term investments. Nevertheless, cost-benefit analyses are often used in coastal adaptation strategies. To resolve the problem of attributing probabilities, it is preferable to use approaches where the uncertainty is represented by a group of scenarios instead of a distribution of probabilities. Robust decision-making is a method characterized by the choice of options considered to be efficient in a broad range of socio-economic and climate scenarios.

The decision-making systems should moreover be flexible so that the decisions can be adjusted if new information and knowledge arises on the climate evolution and on the resulting vulnerabilities of natural and human coastal systems [HAL 09]. The multiple-criteria decision analysis enable robustness and flexibility to be included simultaneously. The application of this type of approach in decision-making processes is becoming increasingly frequent.

The institutional and governmental framework that supports the coastal zones' adaptation is a very important issue because it addresses a particularly dynamic and iterative process where various governmental structures interact at different levels, local, regional and national. The decisions arising from the public and private sectors should take into account different competing and sometimes contradictory interests. For instance, questions of governance become very significant in retreat strategies because the associated political costs are in general very high. The efficiency of adaptation processes increases if the system of governance assures the horizontal and vertical integration of decision-making.

4.6.3. *Adaptation in practice*

The adaptation of coastal zones to climate change is a fairly recent process that is not isolated but which should integrate and contribute to completing the sustainable development strategies for coastal zones that are underway. A particularly favorable strategic framework is the integrated coastal zones management (ICZM), a concept defined at the United Nations Conference on Environment and Development at Rio de Janeiro in 1992 and the Johannesburg World Summit on Sustainable Development in 2002. This strategy will be detailed and broadly discussed in the following volumes of the Seas & Oceans collection. ICZM is a dynamic, iterative, institutional and long-term process of management, integrating the various coastal activities, socio-economic sectors and stakeholders – institutions, non-governmental organizations, coastal communities, business players and science experts – with the objective of using the coastal resources in a sustainable way. The integration of adaptation to climate change in the ICZM is facilitated by its institutional framework, which involves diverse socio-economic sectors, and by its long-term perspective. The main difference in the adaptation process when compared to the ICZM is the greater uncertainty and the much broader time horizons [TOB 10].

The implementation of the ICZM worldwide has proved difficult and its successes are still limited. In Europe, the main difficulties result from the complexity and diversity of the laws governing the coastal zones, from the absence of sustained funding in several countries and from the low participation of stakeholders, business players and citizens in the integrated management process [SHI 07].

For countries in the developing regions the main obstacles are limited knowledge and scientific research, the lack of qualified human resources, institutional capacities, horizontal and vertical cooperation between governmental bodies and the rarity of long-term finance.

Another approach that provides an alternative framework for adaptation, and which has already been applied to the management of coastal zones, is adaptive management (AM) [WAL 97]. This methodology has been developed in the context of ecosystem management to face the uncertainties resulting from the difficulties in predicting the outcomes of management interventions. AM is an iterative and systematic process for the improvement of management practices based on lessons taken from earlier practices. It begins by formulating a management hypothesis that is later tested by implementing it, monitoring the outcomes and learning from these to refine the initial hypothesis. There are strong constraints on the application of AM for coastal adaptation that result from the potential high cost and difficulties associated with experimentation in the context of the impacts and vulnerabilities of coastal zones to climate change.

Community Based Adaptation (CBA) is an adaptation strategy at a local level whose objective is to respond to the impacts of climate change and also to reinforce in an integrated perspective the adaptive capacity of poor communities that are vulnerable to climate and non-climate risks [REI 09]. The aim of CBA is to promote the active participation of different players in the coastal communities of developing countries in a learning process of good social, economic, institutional and environmental practices for sustainable development. This process primarily involves the questions regarding water resources, food security and the adaptive capacity to the impacts of climate change on coastal zones, particularly those associated with extreme weather events and GMSL rise. The International Institute for Sustainable Development (www.iisd.org) is one of the international institutions particularly active on the CBA.

The success of the adaptation process of coastal zones to climate change depends to a large extent on the efficiency of communicating the results of scientific research to political decision-makers and stakeholders at a local level. It is also very important that the decision-making process benefits from the active, integrated and inclusive participation of a broad spectrum of stakeholders, coastal institutions and communities [SCH 13].

4.6.4. *The costs of adaptation*

Even in cases where an adaptation strategy is developed and the corresponding measures are applied with success, it is very likely that it will be impossible to prevent entirely the negative impacts of climate change on the coastal zones. This means that when calculating the cost of climate change, the adaptation costs must be added to the cost of the residual impacts, once the adaptation measures are implemented. As far as the adaptation costs are concerned, they result from the investment costs and the costs of implementing the adaptation measures [FAN 10]. The economic efficiency of an adaptation strategy depends upon a good distribution between adaptation and residual costs that minimizes the global cost. Adaptation to climate change supported by the private sector, that is to say by families and businesses, is in general insufficient, especially in the medium and long-term. Coastal zones adaptation usually requires the intervention of the public sector and especially the cooperation between the various stakeholders concerned, both public and private, at the local, regional and national levels. Adaptation also involves numerous institutions and organizations at different levels. Therefore the adaptation initiatives taken by various institutions and organizations should be coordinated and not contradictory, in order to avoid unnecessary costs. This question is particularly important when defining responsibilities on different spatial scales to coordinate regional territorial policy decisions.

The projects FUND [ANT 10] and DIVA [HIN 09] are the main sources of information regarding the estimated adaptation costs for coastal zones at a global level. One of the most important conclusions of these studies is that the costs of adapting the coasts (adaptation plus residual costs) to GMSL rise during the 21st Century, on a global scale, are much smaller than the costs of the impacts that would result from the increase in the risk of flooding and erosion in the absence of adaptation. Even considering the most disastrous scenarios for GMSL rise by the end of the century, it is economically preferable to protect the higher risk regions of the coast with dykes and other hard structures [NIC 06, NIC 08] than to do nothing. The estimated economic advantage of adaptation – defined as the difference between the costs of the impacts without adaptation and the sum of the adaptation costs plus the costs of the residual impacts – in Europe, in 2020, varies between 3.8 and 4.2 billion euros [RIC 10]. The same models indicate that these values increase as we approach the end of the century.

The costs associated with the impacts of GMSL rise and with adaptation are very variable depending on the country and the regions. Some developing countries, such as for example small economically developing islands with low elevation, have a very limited capacity to support the costs of implementing an adaptation strategy and the costs of the residual impacts. The projected costs of adaptation and residual damage, for a GMSL rise between 0.6 m and 1.3 m until the end of the century, are between 1% and 9% of GDP in developing countries with coastal zones vulnerable to flooding and submersion [ANT 10, HIN 13]. Some of the most vulnerable countries are the low elevation small island developing states in the tropical regions of the Pacific Ocean (Salomon Islands, Marshall Islands, Kiribati, Tonga, Tuvalu and Vanuatu), the Indian Ocean (Maldives and Seychelles) and the Caribbean Sea (Antigua, Barbados and Nevis) [ANT 10]. China, India, Bangladesh, Thailand, Myanmar, Vietnam, the Philippines, Mozambique, Egypt and Nigeria are also developing countries with coastal zones that are particularly vulnerable to GMSL rise. It is likely that some of the most vulnerable countries, especially the small island developing states, will reach the limit of their capacity to adapt before the end of the century. In this situation, it will be necessary to have recourse to retreat strategies, to the planned migration of populations to other countries or to strategies involving the construction of protection structures with outside aid. A World Bank study on the adaptation costs in countries in developing regions [WOR 10] concludes that the main component of the adaptation costs comes from the construction and maintenance of dykes for coastal and river protection in populated areas close to the coast, followed by the costs of beach nourishment and, lastly, the costs of protecting port infrastructures.

Developed countries have a greater capacity to adapt to the impacts of climate change on their coastal zones, but the value of coastal assets subject to the risk of flooding and submersion is considerably higher than in developing countries. It is very likely that from now until the end of the century, the increase in adaptation costs will make the protection of all coastal assets impossible in many developed countries. It therefore becomes necessary to make choices and prioritize the assets for protection.

The vast majority of adaptation cost estimates have been made from a strategic point of view of protection, and mostly with hard engineering structures. It is more difficult to make cost estimates for accommodation strategies, due to the greater diversity of adaptation options and the greater variability of costs in different regions of the world. In the future, it is particularly important to develop further, the cost estimates of facilitating the adjustment of the ecosystems themselves since it is likely that the conservation of natural coastal systems constitutes an adaptation approach with more chance of being successful in the medium and long-term. It is also necessary to bear in mind that, for highly urbanized low elevation coastal regions, vulnerable to the risk of flooding and submersion, the acceleration in the annual rate of increase of GMSL will lead to situations where hard coastal protection infrastructures can be seriously damaged or ruptured, which could be catastrophic.

There are still few studies concerning the adaptation of coastal zones in multisecular time horizons where the rise in GMSL could reach several meters. In Tol *et al.* [TOL 06], we can find an analysis and discussion on adaptation in three European regions – the deltas of the Rhône and Rhine and the City of London – to an extreme scenario where, starting in 2030, the GMSL rises five meters during a period of about 100 years, due to the collapse of the West Antarctic Ice Sheet. In these examples, the likely adaptation strategies for responding to the risk of flooding and submersion vary between the retreat from highly vulnerable urban zones, accommodation and protection with very large and high dykes. The situation will be even more difficult in developing countries since the lower adaptive capacity increases considerably their vulnerability in such extreme scenarios.

4.7. Bibliography

[ANT 10] ANTHOFF D., NICHOLLS R.J., TOL R.S.J., "The economic impact of substantial sea-level rise", *Mitigation and Adaptation Strategies for Global Change*, vol. 15, no. 4, pp. 321–335, 2010.

[CAL 13] CALAFAT F.M., CHAMBERS D.P., "Quantifying recent acceleration in sea level unrelated to internal climate variability", *Geophysical Research Letters*, vol. 40, pp. 3661–3666, 2013.

[DON 07] DONEY S., *et al.*, "Impact of anthropogenic atmospheric nitrogen and sulphur deposition on ocean acidification and the inorganic carbon system", *Proceedings of the National Academy of Sciences of the United States of America*, vol. 104, no. 37, pp. 14580–14585, 2007.

[CHA 08] CHAO B.F., WU Y.H., LI Y.S., "Impact of artificial reservoir water impoundment on global sea level", *Science*, vol. 320, pp. 212–214, 2008.

[CHU 11] CHURCH J., WHITE N., "Sea-level rise from the late 19th to the early 21st century", *Surveys in Geophysics*, vol. 32, pp. 585–602, 2011.

[DIA 08] DIAZ R.J., ROSENBERG R., "Spreading dead zones and consequences for marine ecosystems", *Science*, vol. 321, no. 5891, pp. 926–929, 2008.

[FAN 10] FANKHAUSER S., "*The costs of adaptation*", *Wiley Interdisciplinary Reviews: Climate Change*, vol. 1, no. 1, pp. 23–30, 2010.

[GAP 06] GAP, Global Program of Action for the Protection of the Marine Environment from Land-based Activities, UNEP, 2006.

[GAR 11] GAR, Global Assessment report on Disaster Risk Reduction 2011: Revealing Risk, Redefining Development, UNISDR, 2011.

[GAT 13] GATTUSO J.P., MACH K.J., MORGAN G.M., "Ocean acidification and its impacts: an expert survey", *Climatic Change*, vol. 117, pp. 725–738, 2013.

[GRI 10] GRINSTED A., MOORE J.C., JEVREJEVA S., "Reconstructing sea level from paleo and projected temperatures 200 to 2100 AD", *Climate Dynamics*, vol. 34, pp. 461–472, 2010.

[GRI 12] GRINSTED A., MOORE J.C., JEVREJEVA S., "Homogeneous record of Atlantic hurricane surge threat since 1923", *Proceedings of the National Academy of Sciences USA,* vol. 109, no. 48, pp. 19601–19605, 2012.

[GRI 13] GRINSTED A., MOORE J.C., JEVREJEVA S., "Projected Atlantic hurricane surge threat from rising temperatures", *Proceedings of the National Academy of Sciences USA*, vol. 110, no. 14, pp. 5369–5373, 2013.

[HAL 09] HALLEGATTE S., "Strategies to adapt to an uncertain climate change", *Global Environmental Change-Human and Policy Dimensions*, vol. 19, no. 2, pp. 240–247, 2009.

[HAL 13] HALLEGATTE S., GREEN C., NICHOLLS R.J., *et al.*, "Future flood losses in major coastal cities", *Nature Climate Change*, vol. 3, pp. 802–806, 2013.

[HAN 11] HANSON S., NICHOLLS R., RANGER N., *et al.*, "A global ranking of port cities with high exposure to climate extremes", *Climatic Change*, vol. 104, no. 1, pp. 89–111, 2011.

[HIN 09] HINKEL J., KLEIN R.J.T., "The DINAS-COAST project: developing a tool for the dynamic and interactive assessment of coastal vulnerability", *Global Environmental Change*, vol. 19, no. 3, pp. 384–395, 2009.

[HIN 13] HINKEL J., VAN VUUREN D.P., NICHOLLS R.J., *et al.*, "The effects of mitigation and adaptation on coastal impacts in the 21st century", *Climatic Change*, vol. 117, pp. 783–794, 2013.

[HOP 12] HOPKINS T.S., BAILLY D., ELMGREN R., *et al.*, A systems approach framework for the transition to sustainable development: potential value based on coastal experiments", *Ecology and Society*, vol. 17, no. 3, p. 39, 2012. Available at: http://dx.doi.org/10.5751/ES-05266-170339.

[IPC 13] IPCC, Working Group I Contribution to the IPCC Fifth Assessment Report, Climate Change 2013: The Physical Science Basis, Summary for Policymakers, 2013.

[JON 12] JONGMAN B., WARD P.J., AERTS J.C.J.H., "Global exposure to river and coastal flooding: long term trends and changes", *Global Environmental Change-Human and Policy Dimensions*, vol. 22, no. 4, pp. 823–835, 2012.

[JEV 10] JEVREJEVA S., MOORE J.C., GRINSTED A., "How will sea level respond to changes in natural and anthropogenic forcings by 2100?", *Geophysical Research Letters*, vol. 37, p. L07703, 2010.

[KEM 11] KEMP A.C., *et al.*, "Climate related sea-level variations over the past two millennia", *Proceedings of the National Academy of Sciences USA*, vol. 108, no. 27, pp. 11017–22, 2011.

[KNU 10] KNUTSON T.R., *et al.*, "Tropical cyclones and climate change", *Nature Geoscience*, vol. 3, pp. 157–163, 2010.

[KUL 12] KUHLBRODT T., GREGORY J.M., "Ocean heat uptake and its consequences for the magnitude of sea level rise and climate change", *Geophysical Research Letters*, vol. 39, no. L18608, 2012.

[LEV 13] LEVERMANN A., *et al.*, "The multimillennial sea-level commitment of global warming", *Proceedings of the National Academy of Sciences USA*, vol. 110, no. 34, pp. 13745–13750, 2013.

[LEM 01] LEMPERT R., SCHLESINGER M.E., "Climate-change strategy needs to be robust", *Nature*, vol. 412, no. 6845, p. 375, 2001.

[MCG 07] MCGRANAHAN G., BALK D., ANDERSON B., "The rising tide: assessing the risks of climate change and human settlements in low elevation coastal zones", *Environment and Urbanization*, vol. 19, pp. 17–37, 2007.

[MOS 10] MOSS R.H., *et al.*, "The next generation of scenarios for climate change research and assessment", *Nature*, vol. 463, no. 4782, pp. 747–756, 2010.

[NER 10] NEREM R.S., CHAMBERS D.P., CHOE C., *et al.*, "Estimating mean sea level change from the 35 TOPEX and Jason Altimeter missions", *Marine Geodesy*, vol. 33, pp. 435–446, 2010.

[NEW 12] NEWTON A., "A systems approach for sustainable development in coastal zones", *Ecology and Society*, vol. 17, no. 3, p. 41, 2012.

[NIC 06] NICHOLLS R.J., TOL R.S.J., "Impacts and responses to sea-level rise: a global analysis of the SRES scenarios over the twenty-first century", *Philosophical Transactions of the Royal Society A: Mathematical, Physical and Engineering Sciences*, vol. 364, no. 1841, pp. 1073–1095, 2006.

[NIC 08] NICHOLLS R.J., HANSON S., HERWEIJER C., *et al.*, Ranking Port Cities with High Exposure and Vulnerability to Climate Extremes, OECD Environment Working Papers, no. 1, OECD Publishing, 2008.

[PEA 13] PEARCE A.F., FENG M., "The rise and fall of the "marine heat wave" off Western Australia during the summer of 2010/11", *Journal of Marine Systems*, vol. 111–112, pp. 139–156, 2013.

[POL 13] POLOCZANSKA E.S., *et al.*, "Global imprint of climate change on marine life", *Nature Climate Change*, vol. 3, pp. 919–925, 2013.

[RAH 07] RAHMSTORF S., "A semi-empirical approach to projecting future sea-level rise", *Science*, vol. 315, pp. 368–370, 2007.

[REI 09] REID H., CANNON T., BERGER R., *et al.*, (eds.), *Community Based Adaptation to Climate Change*, International Institute for Environment and Development, London, 2009.

[RIC 10] RICHARDS J.A., NICHOLLS R.J., Impacts of climate change in coastal systems in Europe. PESETA – Coastal Systems Study, JRC Scientific and Technical Reports, European Commission, Joint Research Center, 2010. Available at: http://ftp.jrc.es/EURdoc/JRC55390.pdf.

[RID 10] RIDLEY J., GREGORY J.M., HUYBRECHTS P., *et al.*, "Thresholds for irreversible decline of the Greenland ice sheet", *Climate Dynamics*, vol. 35, pp. 1065–1073, 2010.

[SAN 02] SANTOS F.D., FORBES K., MOITA R., "Climate change in Portugal. Scenarios, impacts and adaptation measures", SIAM Project, Gradiva, Lisbonne, 2002. Available at: http://siam.fc.ul.pt/siamI_pdf/.

[SHE 12] SHEPHERD A., *et al.*, "A reconciled estimate of ice-sheet mass balance", *Science*, vol. 338, pp. 1183–1189, 2012.

[SHI 07] SHIPMAN B., STOJANOVIC T., "Facts, fictions, and failures of integrated coastal zone management in Europe", *Coastal Management*, vol. 35, no. 2, pp. 375–398, 2007.

[SCH 13] SCHMIDT L., DELICADO A., GOMES C., *et al.*, "Change in the way we live and plan the coast: stakeholders discussions on future scenarios and adaptive strategies", in CONLEY D.C., *et al.*, (eds.), *Journal of Coastal Research*, no. 65, pp. 1033–1038, 2013.

[SRE 00] SRES, Special Report on Emissions Scenarios: A special report of Working Group III of the Intergovernmental Panel on Climate Change, in NAKICENOVIC N., SWART R. (eds.), Cambridge University Press, Cambridge, 2000.

[STE 04] STEFFEN W., *et al.*, *Global Change and the Earth System: A Planet under Pressure*, Springer-Verlag, Berlin, 2004.

[TOB 10] TOBEY J., RUBINOFF P., ROBADUE JR. D., *et al.*, "Practicing coastal adaptation to climate change: lessons from integrated coastal management", *Coastal Management*, vol. 38, pp. 317–335, 2010.

[TOL 06] TOL R.S.J., *et al.*, "Adaptation to five meters of sea level rise", *Journal of Risk Research*, vol. 9, no. 5, pp. 467–482, 2006.

[UNE 11] UNESCO, in PIGUET E., *et al.*, (eds.), *Migration and Climate Change*, UNESCO Publishing and Cambridge University Press, 2011.

[VAF 11] VAFEIDIS A., NEUMANN B., ZIMMERMANN J., *et al.*, "MR9: analysis of land area and population in the low-elevation coastal zone (LECZ)", *Review in Commission of the Foresight Project "Migration and Global Environmental Change"*, 2011.

[VER 09] VERMEER M., RAHMSTORF S., "Global sea level linked to global temperature", *Proceedings of the National Academy of Science USA*, vol. 106, pp. 21527–21532, 2009.

[WAL 97] WALTERS C., "Challenges in adaptive management of riparian and coastal ecosystems", *Ecology and Society*, vol. 1, no. 2, 1997.

[WEB 05] WEBSTER P.J., *et al.*, "Changes in cyclone number, duration, and intensity in a warming climate", *Science*, vol. 309, pp. 1844–1846, 2005.

[WOR 10] WORLD BANK, Economics of Coastal Zone Adaptation to Climate Change, Discussion Paper, no. 10, 2010.

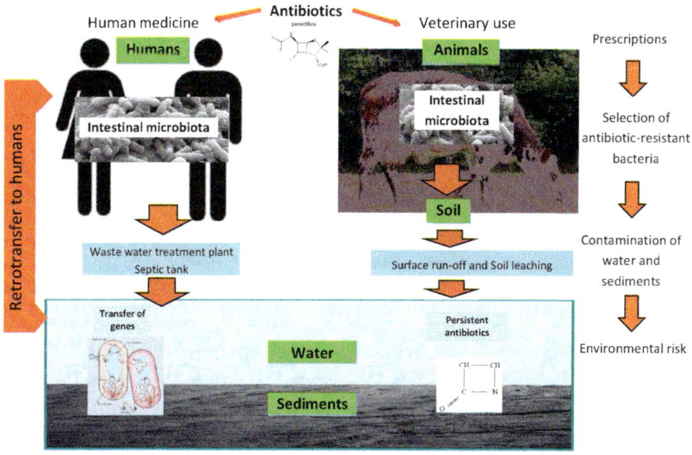

Figure 2.1. *Circulation of antibiotic-resistant microorganisms and antibiotics within the four major ecosystems: human and animals under antibiotic treatment, soil and water*

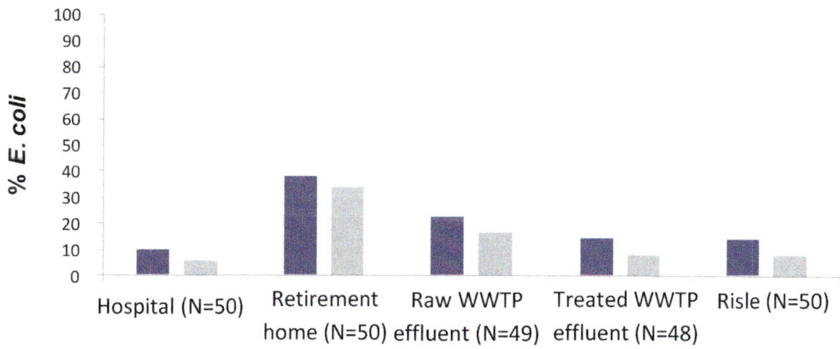

Figure 2.3(a). *Occurrence of multiple-resistant strains of E.coli along the medical centre-WWT river continuum: percentage of isolates of E. coli resistant to at least three antibiotics (dark gray), percentage of isolates of E. coli carrying class 1 integron (light gray) medical center – WWTP river [ORB 11]*

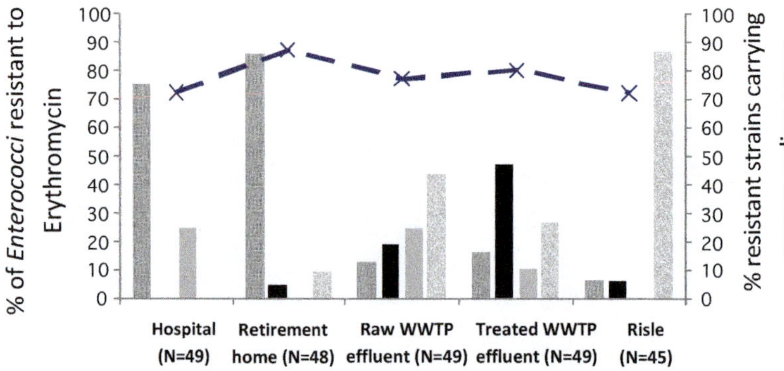

Figure 2.3(b). *Occurrence of Enterococcus strains resistant to erythromycin, along medical center – WWTP – m continuum: (- - -), percentage of Enterococcus strains resistant to erythromycin (macrolides family), percentage of strains of Enterococcus resistant to erythromycin carrying corresponding genes ermB mefA n, ermB+mefA n, unknown genes (Leclercq et al, 2013)*

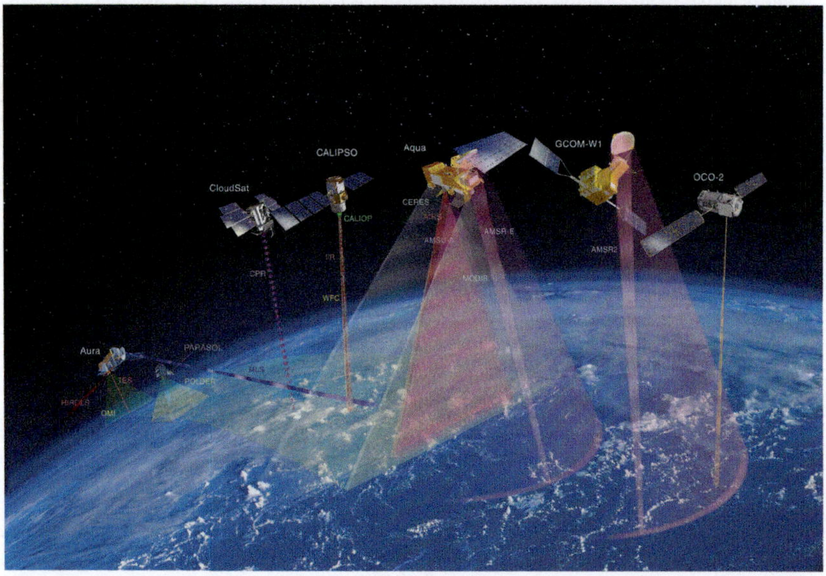

Figure 3.1. *Image of the A-Train carrying satellite AQUA which carries the sensor MODIS used to sense the color of the water (source: NASA)*

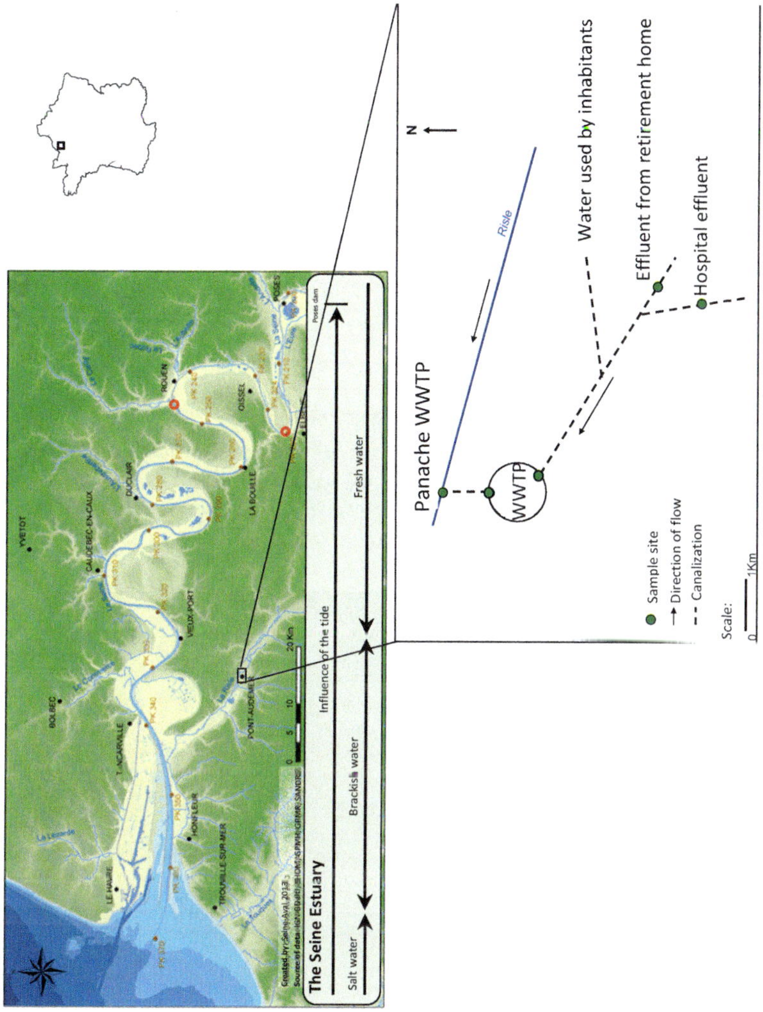

Figure 2.2. *Map of the Seine estuary: one of the most anthropized estuaries in Europe, with a population density stated in inhabitants/km2 (black: 3000-6000 inh/km2; grey 500-3000 inh/km2; white < 500 inh/km2; INSEE dataset), mainly contaminated by WWTP and the upstream input from Poses; Kp, kilometric point (kp 0 at Pont Marie in Paris); and of the studied zone where is located the medical center (hospital of 87 beds and retirement home of 180 beds)-WWTP-river continuum (4km), for which the receiving river corresponds to the tributary (Risle) which flows in the mouth of the Seine estuary*

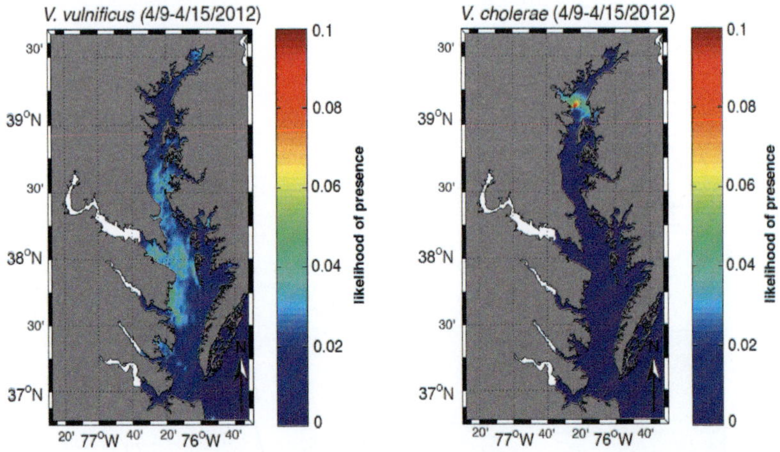

Figure 3.3. *Probability maps for the presence of V. vulnificus and V. cholerae in Chesapeake Bay (United States) generated using interpolated estimations of the surface temperature of the water and the salinity applied to models of the habitat of Vibrio, over the period from 9 to 15 April 2012. These maps have been kindly provided by E. Urquhart (Johns Hopkins University, Baltimore, United States) and R. Colwell (University of Maryland, United States) within the framework of collaboration*

Figure 3.5. *Aerial view of a bloom of cyanobacteria in the Baltic Sea, 13 July 2005 (Source: MERIS-RR (ENVISAT-ESA); credits: ESA)*

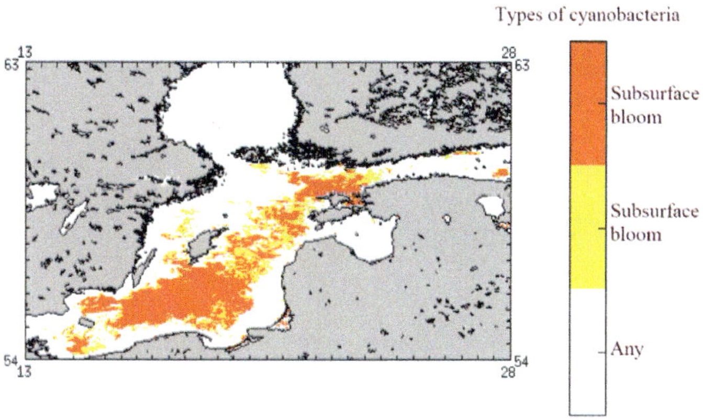

Figure 3.6. *Map of cyanobacteria bloom in the Baltic Sea for the day 11 July 2010 (sensor MERIS (ENVISAT-ESA), L3 MEGS, handling ACRI-ST)*

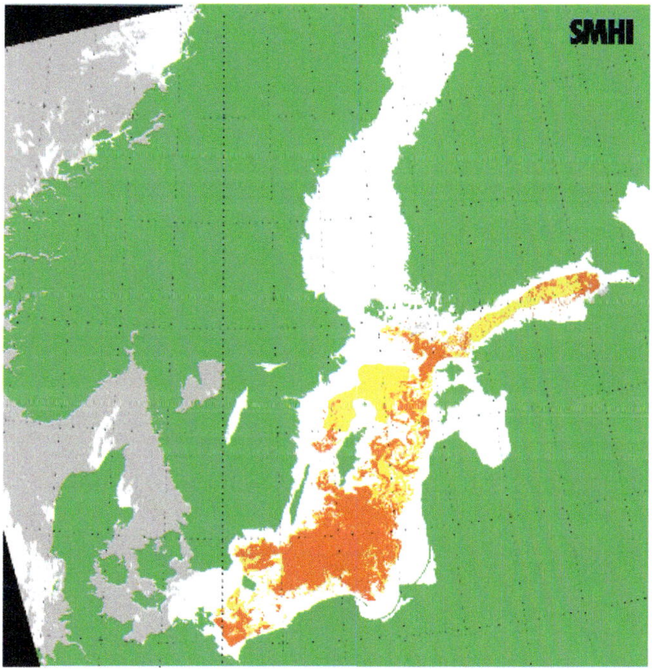

Figure 3.7. *HELCOM bulletin from 11 July 2010 (Source: Swedish Meteorological and Hydrological Institute – SMHI)*

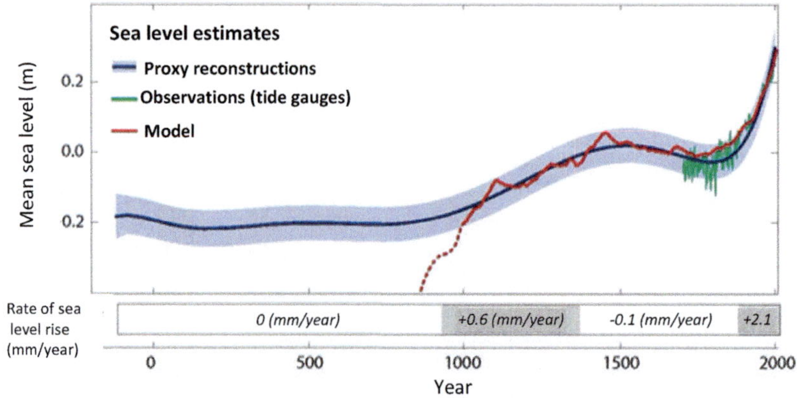

Figure 4.3. *Evolution of GMSL over the last 2,100 years (adapted from [KEM 11])*

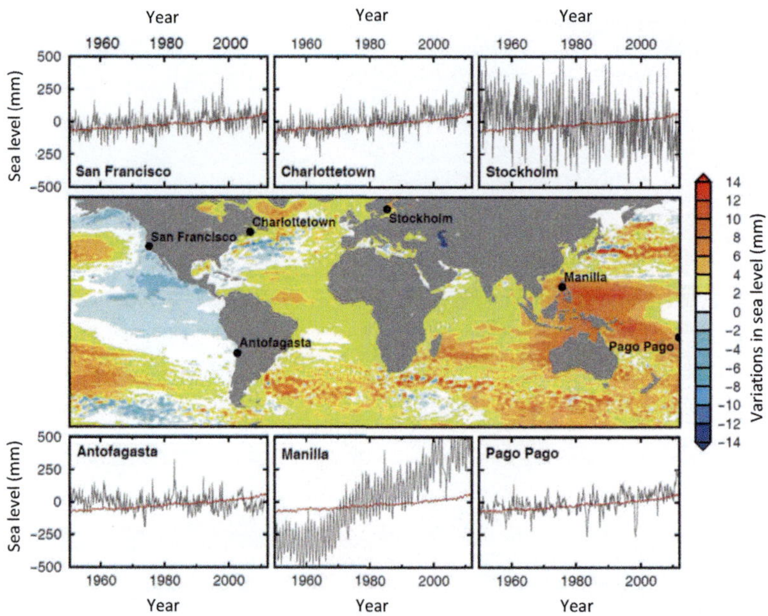

Figure 4.4. *Rates of change of mean sea level in relation to the center of the Earth for the period 1993–2012. The figure also shows the variations (in grey) of the LMSL in six coastal towns, detected by tide gauges, for the period 1950–2012. The lines in red represent an estimate of the variation in GMSL in the same period (adapted from [IPC 13])*

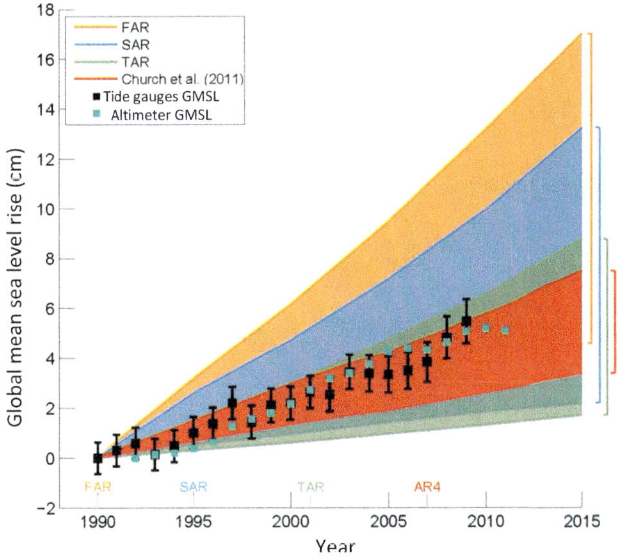

Figure 4.5. *Variation in GMSL from 1990 to 2011, evaluated using data from tide gauges (error bar in black) [CHU 11] and satellites TOPEX and Jason (blue squares) [NER 10]. The colored parts represent the margins in the projections for the GMSL from IPCC reports 1° (FAR), 2o (SAR), 3° (TAR) and 4° (AR4), corresponding to the different scenarios for emissions of greenhouse gas (adapted from [IPC 13])*

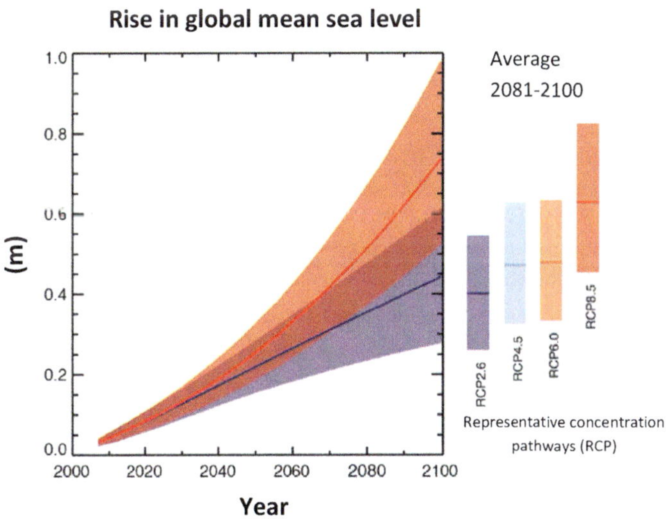

Figure 4.6. *Projections for the rise in GMSL up to the end of the century, resulting from different RCP scenarios [IPC 13]. The margins in blue and red correspond to the uncertainties of the extreme scenarios RCP2.6 and RCP8.5*

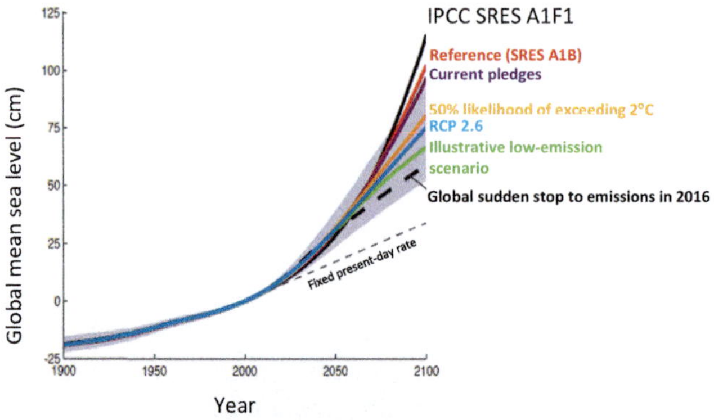

Figure 4.7. *Evolution of GMSL projected by the semi-empirical models with different scenarios for emissions described in the text. According to these models, the current commitments for mitigation are insufficient to avoid a rise in GMSL lower than 50 cm at the end of the century*

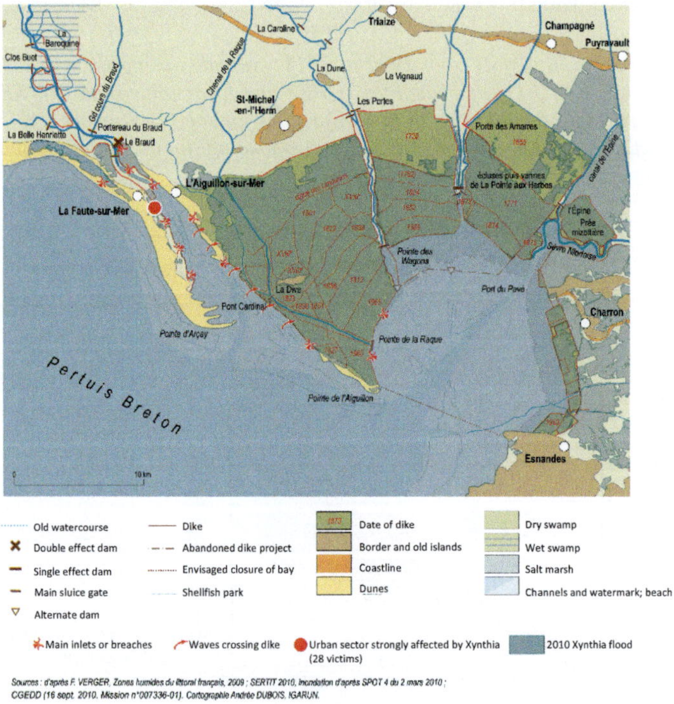

Figure 5.6. *The map shows the magnitude of submersion after the Xynthia storm that almost reached the limits of the first dike for the second half of the 19th Century. The submersion is related to the influx of water and the numerous breaks in the dikes that were poorly maintained or not maintained at all (from [VER 09])*

5

The Shoreline between Nature and Society, a Changing Heritage

5.1. Introduction: the coastal "heritage", a new concept

Although the protection of particular coastal sites can be considered as relatively old, the last four decades have shown a constant effort in that respect on the part of States, dynamic and careful associations and, at the top, the international community. This effort derives from a growing awareness – long-established in the United Kingdom, where the National Trust, founded in 1895, was established to protect monuments and sites, not necessarily situated on the coasts – of the specificity of shorelines, of their originality, both biological and geomorphological, and of their vulnerability as a result of the increased pressure from humans, attracted by the richness of the environment and scenic beauty. This "heritagization" is in line with the primary objectives of the National Trust. In order to ascertain this, let us visit a famous site, the Giant's Causeway in Ulster.

The Giant's Causeway is a natural wonder. The beauty of the site is remarkably well showcased. Isolated at the northern end of the Irish coast and dominating the sea off its high basalt cliffs, the site expresses first of all the beauty of its natural surroundings. The

---

Chapter written by Alain MIOSSEC.

isolation of the site has helped in preserving it from human impacts. The latter can, however, be found all around in the area in typical forms; such as Portrush with its beaches on the west, its golf courses that extend all the way to the shoreline to the south, and finally Ballentoy further east. The majestic basalt promontory dominates these features from afar. The road leading to the Giant's Causeway is narrow and ends in a parking lot of reasonable size, as demure as it can be. The walker then enters the area of the Giant's Causeway: a low building with large windows welcomes the visitors to the right, providing leaflets with information on the site and souvenirs. A mandatory element of the leisure civilization, where the worst can be found alongside the best. Finally to the left, in a long room partitioned by panels, the show begins: geological cross-sections and historic reconstructions (up until the Catholic rebels). This is called making good use of the natural beauty: the exhibition can be visited before walking through the site, or it can be done after a long walk to better seize the interest of what is being shown. Each visitor is free to choose. We are free to watch a movie where fantasy overlaps with realism. The owners of the site do not impose any order, but do offer, in the religious silence of a museum, "an area of outstanding scenic beauty", a crowning jewel of man in the United Kingdom and in the world, the Giant's Causeway. After the exhibition, walkers can follow the paths and enjoy nature. The bravest walkers can take the whole tour while visitors who are in a hurry or tired can choose shorter options, such as the path leading to basalt sea-polished slabs. It is recommended to use the brochure that helps in recognizing Hamilton's Seat and Chimney Tops; with a bit of luck, a peregrine falcon might be spotted among the 50 species that occupy the site. The National Trust welcomes each year several hundreds of thousands of visitors. Everything is clean, well organized, didactical and dynamic. Protecting and promoting the coastline, and designing it for that purpose are truly Anglo-Saxon traditions. It is not enough to open the coast to the public, but explanations are provided as to why we should be proud of this heritage, why the latter should be respected and, finally, why we should leave all the wiser. This is an inspiring example of coastal natural heritage management.

This long preface situates this chapter within the ambivalence of environments and landscapes and in the choices of societies that here decide to open a space up to the public, where other societies would rather fence it off. This situation arises as on this narrow strip of shoreline, conflicting interests clash and the result can be brutal for the environment in question. Between degradation and promotion, the game is on and the stakes are high in terms of protection, conservation and restoration. It is the outcome of a long story, sometimes guided by a top-down approach and taken over by a bottom-up strategy on the ground or, on the contrary, inspired by a range of local initiatives, which then have to be coordinated. It is the scene of diverging logics, as the societies involved are different, even, and especially, in a globalized world whether they operate through centralized management or in the context of a more rooted decision-making process. On the one hand, we have the French tradition and, on the other hand, the federal State model of the United States, two models that are more similar than we tend to think. They have been developed in parallel for 40 years through, on the one side, the implementation of the Coastal Management Act, and, on the other side, a very prescriptive approach which, over time, shows the perseverance of the State in spite of recurring criticisms. These are the stakes in a world where people worry too much about their seas and coasts and develop a pessimistic vision of their future.

Implementing policies for the protection of coastal (and marine) environments is merely the consequence of a triple realization. On the one hand, the growing pressure of man on coastal zones, and on the other hand, the need for a healthier and more sustainable resource management, as we would term it after the Earth Summit in Rio de Janeiro in 1992, that is not only spatial (the coastal space), but also biological, due to the high productivity of interface or coastal environments. Finally, it is also the realization of the fragility of the coastline itself subject to natural hazards, and coastal erosion, which can be insignificant in some places and considerable in others that are subject to storms. The media coverage of these exceptional catastrophic phenomena also helps in managing, in a more intelligent way, this coastal strip exposed for millennia to the action of the waves. It has also made us realize in the past few centuries that getting too close to the sea is not risk-free especially given the sea level rise we have observed for more than a century.

With various uses of a fragile and highly dynamic space and with growing pressure, competing human activities of all types arise. Can we regulate without prohibiting and, if prohibiting is unavoidable, can we do so with the consent of all the stakeholder groups? In other words, can we protect coastal human settlements and the coastal environments from human covetousness? It is a major challenge that requires an integrated management approach of coastal zones.

5.2. Coastal dynamics

The more settlements have moved to coastal areas, the more sensitive these areas have become to changes in the coastline. While the postglacial sea level rise was coming to an end, the residents of the Gulf of Morbihan had to move their settlements further inland, which we are reminded of by a few megaliths immersed by a several meters depth. Nowadays, the same observations are made everywhere, especially as the exposed settlements are numerous and as the coast had to be barricaded by flood defense walls to protect its residents from the waves. Coastal erosion is a natural process: a cliff is a perfect vestige of erosion, and this movement is irreversible. It can be seen each year between Criel-sur-Mer and the Somme Bay, when comparing it with photographs taken successively since the end of the 19th Century that, roll after roll, dozens of meters have fallen into the sea. However, in spite of this observation, things are more complicated as they have developed within a real system that should be explained.

5.2.1. *The morphogenic coastal system*

In this system, five factors play a role [MIO 09] (Figure 5.1).

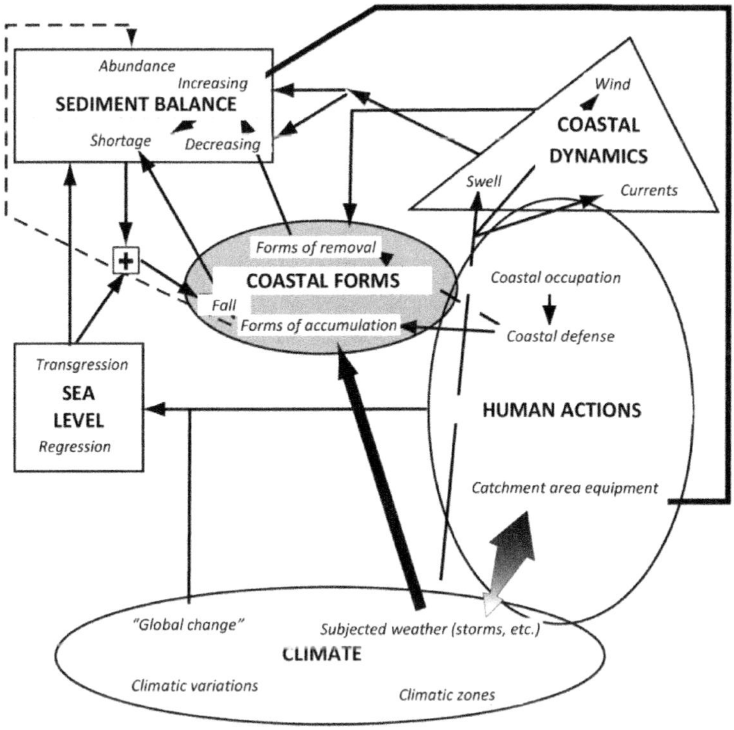

Figure 5.1. *This model links together the interactions involved on the basis of five factors, climate (in all the different meanings of the term), sea level, coastal dynamics as related to climate and the sediment balance and partly related to coastal dynamics and partly to human actions*

The driver of the system is the climate, and it can be interpreted in the wider sense at all the scales, both temporal and spatial. It includes the weather as much as the climatic zone; the recent time-frame (week, year, decade, etc.) as much as the great climatic oscillations that have marked the Quaternary era and of which the coasts still bear the mark (marine sediments reveal the movements of the sea; masses of sediments washed off from the continents and sorted, reworked during marine transgressions). On the coasts, it becomes clear that the first effect of the climate is the wind, sometimes acting through the sea (swells, waves, etc.), and on the coast itself, through the deflation of sandy shores. The first dunes in line manifest the wind by their shapes. Surges gather sediments and wash them laterally (coastal drift) or transversally onto the "profile" of the beaches.

The sediment balance is an essential piece of information whose effects can be taken into account more effectively. It expresses the ratio between the amount of existing sediment in a given location and the ability of the marine agents (the wind, waves and currents) to mobilize them. The forms of accumulation express this balance at all times: in mid-latitudes, the beaches lose sand during winter storms (a simple example) and the long swells at more quiet times bring them back to the coast. The origin of the sediments is a crucial element. The catchment area is the primary source: sediments are poured into the sea through streams and rivers as they are the result of continental erosion. In estuaries and deltas, an estimate of the input can be obtained; in tropical countries, streams bring sand and clays in abundance, whereas at high latitudes, they wash a multitude of sediments that are often poorly sorted. The sea itself is a source of inputs, either by its permanent onslaught on the shores with cliffs and mobile coastlines, or by producing biogenic sediments such as with coral reefs. The swells that batter the continental shelves dislodge the sand down to such depths that scientific debates and research are still struggling to establish a "limit to the effect of swells", which has been estimated around isobaths of 20 m, not without discussion, as the local conditions may vary, and they play a considerable role in the matter. Research in the last decades has shown a degradation of the continental shelves battered by large oceanic swells [BIR 85, PAS 98]: significant examples have been given such as major barrier islands on the Atlantic coast of the United States where, after long periods of accumulation of sand, which is a sign of advancing coastline, the latter has retreated regularly, sometimes since the 16th Century (Figure 5.2). Although this theory is subject to debate, it expresses an element which the engineering behind coastline protection should take into account and which contemporary policies fighting against coastal flooding cannot ignore.

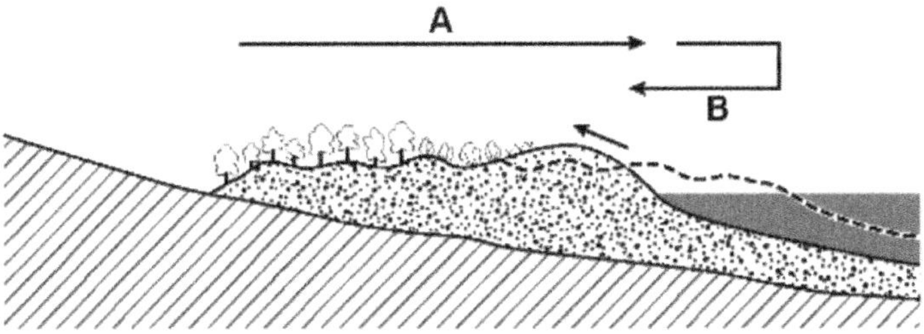

Figure 5.2. *Theoretical evolution of a beach over the past millennia: accumulation of sand as a result of transgression by successive peaks a) start of the retreat b) by onslaught of the sea due to the depletion of under-water resources [BIR 85]*

These dynamic processes generate the third factor: the surges contribute, in a most efficient way, to the phenomenon of coastal morphogenesis. The waves batter the sediments, hit the cliffs, dislodge matter and transport it through longshore currents which are, out of all the existing currents, the most effective and spectacular: open-edged sandspits are their most representative illustration. These processes can be studied at all scales but to create useful models that are relevant to today's issues, these models should rely on data collected on the ground. Specialized journals (*Marine Geology, Journal of Coastal Research,* etc.) provide a good idea of the current research being carried out on the surf zone in terms of sediment mobility related to the action of different types of waves and currents (tidal, rip tides, etc.). These use technical tools that get immersed in the sea and capture the required parameters to better understand the mechanisms in action (current meters, sediment traps, etc.). These studies are important as they are related to experiments aiming to counter the effects of erosion, in particular along the large sandy shores that are subject to tourism development.

Human development, in all its forms, has impacts at all the geographical scales, most of these impacts being restrictive. If possible, for now at least, let us put aside the consequences of human activity at the ocean level (steep rise in the greenhouse effect). The development of the watersheds has greatly reduced the migration of sediments; on the contrary, the clearing waves, even if temporarily, have been able to speed up the washing of the finest matter into the sea. From this point of view, the development of marshes throughout history is probably responsible as much for these abundant inputs as for the management deployed by men to develop these lands (Bay of Mont Saint-Michel but especially the marshes of the center-Atlantic in France or the establishments on large salt marshes that could not have happened in a context of a shortage in sediments; the reclamation of land in the Netherlands, or the coastlines along Korea, China and especially Japan, etc.). Increasingly, the sustained secular and complex watercourse management effort has affected the input of sediments: the placement of dikes has the effect of hardening the shore and prevents sand and finer matter from being washed into streams, whereas the implementation of spikes across streams to facilitate shipping compresses the strength of the current in one part of the riverbed and traps the sediments. Dams, of course, which are increasingly being used, form formidable reservoirs of sediments that are rarely cleaned. On the seaside, the supplying of harbors plays a similar role to spikes but most importantly, since the start of tourism development, numerous stations have required the building of concrete waterfronts and the placing of heavy flood defense structures (various walls, ripraps and stones) which, there again, reduce the input of sediments to beaches. The movement can be endless, the poorly designed spikes (groynes) relay each other and divide up the coastline in so many discrete cells with low permeability, which in turn emphasize the sinking of beaches and the discharge toward shallow sandy stocks. The chronic sediment deficit in many places compels us to rethink our approaches and methods, as the refilling of beaches in various forms only maintains sediment stability temporarily.

Finally, the changes in sea level are key elements of the system but the timescale for these changes is not seasonal or annual. It oscillates within the duration of transgressions and regressions, with a few hiccups. These phenomena deserve particular attention in that they are a more or less quick form of continental submersion.

5.2.2. *Marine submersion*

The story of marine submersion is an ancient one, wrapped in legends that were born from the realities of the moment. The great flood in the Old Testament for instance brings us back to Mesopotamia but also recalls that during the Flandrian transgression, the people who lived not too far from the sea might have had the impression that a catastrophe was taking place, as the threshold of the Black Sea was overcome due to the scale and speed of the phenomenon. Similarly, tales from Brittany of the King of the mythical city of Ys are rooted in similar natural phenomena, assuming that the rhythm of the transgression varied over time with phases of extreme speed. This ancient story periodically comes back to life in the media: the slow sea level rise is threatening some Pacific islands with submersion [BIR 93]. The sea is not the only one at fault as these islands were often utilized in the conflict between the United States and Japan to store air fields whose materials were taken from coral reefs that were greatly undermined in the process. It is also the case in the Maldives, where tourism development also occurs to the detriment of the coral reef, which is the best existing flood protection.

If the risk of slow submersion exists, it is also down to man to do something about it; however, this assumes that we take into account both the history of the latest transgression and the causes that can be identified to explain its rhythm and coastal effects. This is a truly topical question that is echoed by the media in the context of climate change according to some, whereas others, although they do not deny climate change, fail to see the early stages of a real change that would imply instability in the major action centers of the climate system (anticyclones and marine currents).

5.2.2.1. *A bit of history on the latest transgression*

Research on the topic of sea level rise extends along various lines [PIR 96]. All studies highlight a convergence of the sea level rise curves in a transgression maximum around 6,000 BP, but most diverge when it comes to identifying the rhythms (although not significantly) due to the interference of sea movements and continental shelves. The curve shown below is one of the most recent and represents Tahiti, where several measurements have been taken in the coral reef.

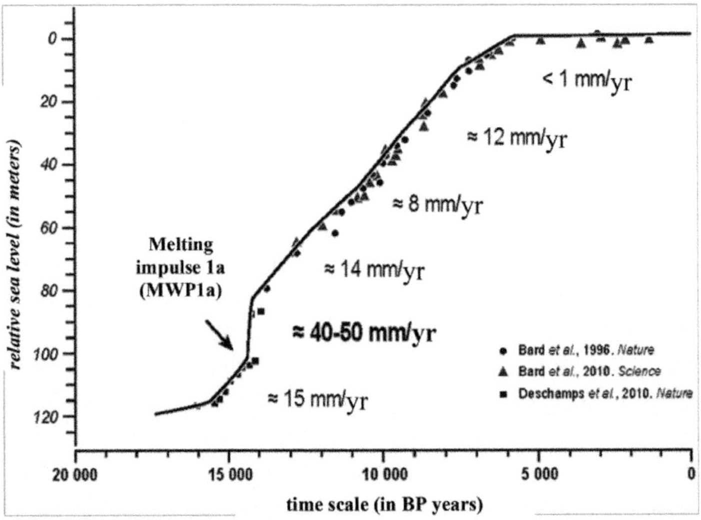

Figure 5.3. *Evolution curve of the marine environment in Tahiti. Dating (U-Th) the corals reveals a spectacular change in the sea level of 120 m in about ten millennia*

As highlighted by most specialists, it is toward 18,000 BP that the level started to rise at a moderate rate. The ocean level was then of 120 m lower than it is at present. Then, things started speeding up, the speed is relevant: 15 mm/year between 14,000 and 16,000 years, then from 40 to 50 mm/year is a few centuries. It is fascinating. From 14,000, the speed decreased sharply but remained high (especially compared to today's rate), between 8 and 14 mm/year until around 6,000 years BP. This rise was related to a warming of the climate that made the great inlandsis of the northern hemisphere retreat. This can still be witnessed today with the retreat of Greenland; however, it is not (yet) comparable with the previous case. The sea level rise, which is rapid if we place it within the geological history of our planet, has had various consequences on the coastlines: immersed shorelines (shingle beaches and mudflats on the Breton precontinent that were marine marshes at a certain point in history) mark the transgression, and the sea has pushed back, within the geomorphological limits of its abilities, with heaps of matter, sand and shingles, and mudstones of course, but whose size allowed for all types of dispersion [PIN 74, VAN 77]. The sandy shores as they stand today are only the outcome of the transgression in question coming to an end with, since then, a constant reprocessing of the matter that tends to correspond to the dominant swells. So that when humans have frozen the shoreline with heavy structures, they have introduced noise parameters in the order of nature. Many bitter experiences on the ground are the consequences of misunderstanding the mechanisms of transgression itself.

At the dawn of the 21st Century, sea level rise is an element of the debate on global warming that cannot be denied. In summary, the temperatures have started to rise since the years 1880, i.e. long before the massive industrial development of the last decades. The causes of it are "natural"; we are at the end of the Little Ice Age, although the coal burning activities of the last 50 years might have played a role in that. The 20th Century has seen a rise in temperatures of the order of 0.5°C between 1880 and 1940, then it went down by 0.2°C until the end of the 1960s. The sea level is thought to have risen by about 15 cm including all the known causes (steric effect, etc.). Should we "fear the worst" as Roland Paskoff wrote in his controversial book [PAS 93]? Surely not in the short term or beyond recent controversies, publicized by the media and so therefore partly curtailed of their scientific accuracy. At the time, nothing indicates that more than a 50-cm rise can be expected during this century; which is disclosed by the latest IPCC report and confirmed by September 2013 report. Indeed, this report gives a fork included between 26 and 82 cm for 2100, which has not changed since the 2007 predictions except for the alarmist tone that puts the emphasis on human responsibilities in the situation. Of course, these curves indicate a relative acceleration (of 3.2 mm/year), taking into account the fluctuations arising due to El Niño and La Niña oscillations. However, it is important to note and discuss the results and the place from which they have been obtained: the coastlines are a human heritage; they are becoming increasingly crowded, and the risks are growing accordingly. Recent catastrophes, the Japanese and Indonesian tsunamis [LAV 11], and the recent storms in the Atlantic, ring so many alarm bells but there is no need to anticipate and attempt to predict future changes, because the risk is very much present.

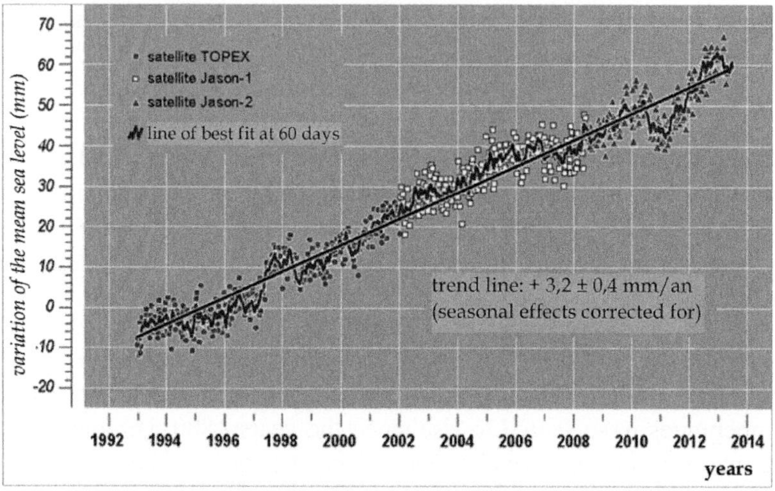

Figure 5.4. *Recent time series of measurements of global sea level (corrected for seasonal effects)(source: Sea-Level Research Group, University of Colorado, 2013)*

5.2.2.2. *The effect of marine meteorological events*

This ancient story also resurfaces when storms have catastrophic effects and, in France, after Xynthia, the State has been aiming to reinforce its ability to respond to such events. On the coasts, these storms punctuate the evolution of the coastline. Storms and typhoons (tropical cyclones), and the even more spectacular and dangerous, albeit less frequent, tsunamis, add pressure to the shorelines. One of the most famous and best studied cyclones, for its functioning as well as for its effects, is Hugo, which swept over the western Atlantic side of the West Indies (Guadeloupe was deeply impacted) to the coasts of the United States, where both Carolinas were particularly affected by it. When it hit South Carolina, the atmospheric pressure in the eye of the hurricane was 934 millibars and wind speeds were at their strongest, 216 km/h. The storm surge was stunning, 4–6 m depending on the location, and the overwash reached more than 100 m inland. The morphological consequences were significant: rupture of the shorelines in coral reef islands (Figure 5.5), with a transfer of water and sediments inland and, at the end the sequence, shorelines became shattered into multiple passages.

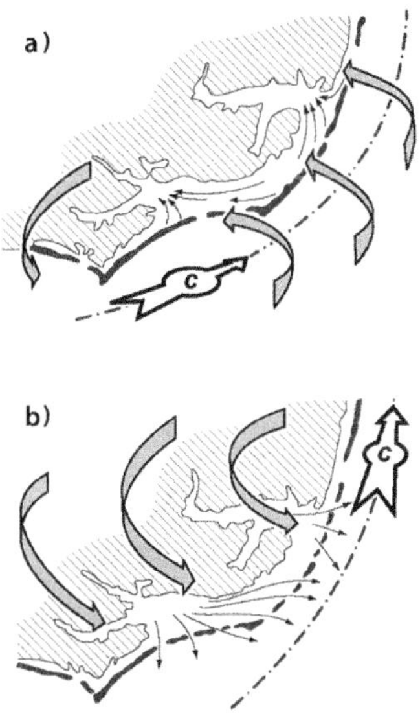

Figure 5.5. *Opening of passages in the "reef-islands" during a hurricane on the south-eastern coast of the United States according to [LEA 79]*

The retreat of dune feet is impressive; it was measured at Myrtle Beach (famous beach resort in South Carolina) to be between 8 and 18 m in the few hours in which the sea was at its wildest. As for the storm Xynthia, the most recent one among those that periodically affect the French Atlantic coastline, it deployed slightly less energy in the model (Figure 5.6). On 27 February 2010, the Atlantic coast was under threat of a violent storm, with four departments under red flags, including the Vendée and Charente Maritime, and it was hit during the night of February 27 and 28: southwesterly winds at 110–120 km/h with peaks at 140–160 km/h in Oléron and Ré. These are not the highest wind speeds recorded (Klaus storm in 2009, more than 200 km/h). The tidal coefficient was high (but far from the theoretical maximum of 120), at 102, and the trough was deep, at 977 hpa. The atmospheric and tidal sea surge increased the levels to up to 4.7 m General Leveling of France (GLF). This was of course an exceptional phenomenon, but it is not rare: marine submersion is recurrent and has been recorded since the end of the Middle Ages, at least in the cases when some settlements were affected and there were a few victims, which were never high in numbers as the coasts were not as densely populated as today. The 29 deaths recorded in La Faute-sur-Mer are only a consequence of the recent and socially reckless establishment in sectors that were affected by past and "forgotten" submersion events. These invoke a lack of realization, a "forgetful" memory so that the responsibilities are widely shared from the top to the bottom of the social pyramid.

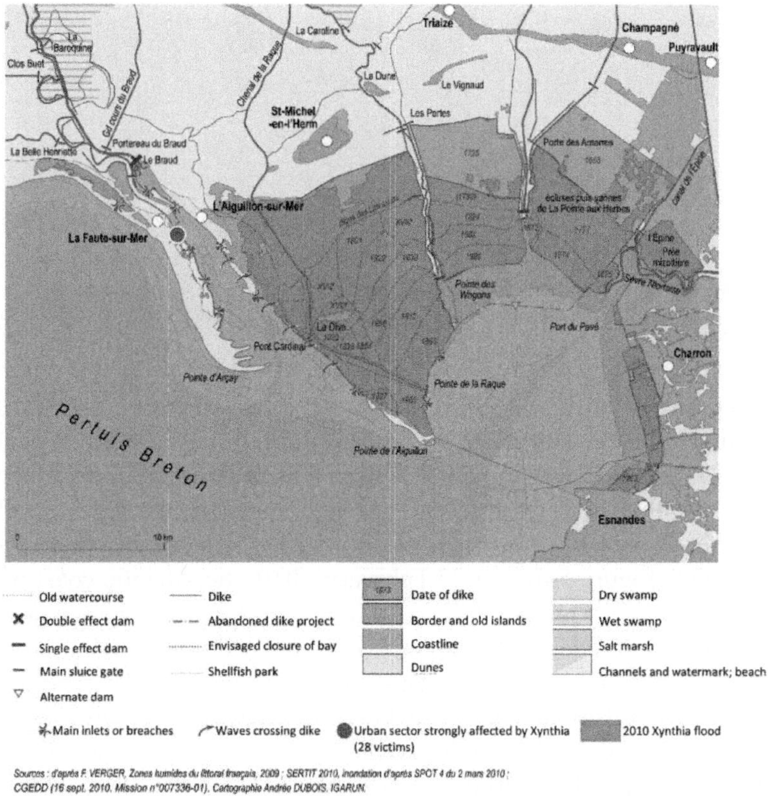

Figure 5.6. *The map shows the magnitude of submersion after the Xynthia storm that almost reached the limits of the first dike for the second half of the 19th Century. The submersion is related to the influx of water and the numerous breaks in the dikes that were poorly maintained or not maintained at all (from [VER 09]) (see color section)*

It is under the constant threat of these supposedly exceptional marine-meteorological conditions, but which in fact are not that uncommon [CAR 10, FEU 12], that the shorelines evolve and retreat. To summarize, all the factors likely to affect the coasts that do not have the potential to modify things need to be combined: deep-low pressures, and therefore strong rotating winds that increase in intensity as the trough advances, so these progressive changes in wind direction further complicate the progression of swells travelling from afar combined with the local action of the wind that results in storm surges. In other words, the surf is much higher than would normally be expected and is also a function of the tidal coefficients. Many violent storms have made little damage to the coast because they developed in neap tides or at a point of the tide where the latter was not at its maximum. Xynthia made a lot of damage as at dawn, the intensity of the winds and the violence of the surges coincided with the maximum day tide. The inflow of water, particularly in confined zones such as estuaries, which provoke the submersion of the beaches, is a natural system, and the breaking of the dikes, that had been built out of necessity long ago, are often poorly maintained (Figure 5.6). The solution – if any is likely to be implemented in all cases – resides in the organization of society and partly – only partly – in scientific research. In the United States, everyone knows how to mobilize populations quickly in the event of a tropical cyclone, and there are generally few victims. The same cannot be said of other places. In France, things only change after there have been victims. A procedure for the prevention of marine submersion and rapid flooding named "dike procedure" was drafted in urgency. It involves, among others, repairing and reinforcing the 1,200 km of dike between 2011 and 2016. This program costs 500 million Euros, of which 40% will be funded by the risk prevention funds (Barnier funds) and should also benefit from European funds (FEDER). The strategy aims to control urbanization in risk areas. On the other hand, another line of the plan aims to improve the warning, monitoring and response chain. To that end, a section on "strong waves – marine submersions" has been added (at the department level) by the weather forecast and has been implemented by Météo France as a new open and regional warning service of "rain flood", whereas each district concerned by risks of submersion and flooding will have to be equipped with a district safeguarding procedure (encouraging within-district approaches).

Reinforcing the flood defense structures involves speeding up the inventory of the dikes and selecting the maintenance work that needs to be done. The ministry in this occasion organized the management of dike structures. This is neither new nor easy to implement, as carrying out an inventory of these structures is all the more delicate as their management depends on various collectives. Finally, the "dike procedure" anticipates a better understanding and a change in the risk culture in terms of marine submersion. Things are progressing slowly, submersion percolates with difficulty within the SCOT, for instance, yet its scale can be relevant. Consultations are difficult to put into practice; this is another way that management could be carried out more efficiently in the perspective of a more sustainable development.

5.3. Anthropogenic pressures: coastlines in the face of human pressures

5.3.1. *Demographic growth expresses the increasing attraction of coasts*

When looking at a global demographic map, we notice the concentration of a large portion of the population on the coasts. This observation is derived from historical migrations of populations and was emphasized during the second half of the 20th Century.

Historical migrations are the first and foremost factor in population increases among the coasts and in particular during the 16th Century (see Figure 5.7(a)), and up until the 20th Century (see Figure 5.7(b)), due to the European expansion toward America [GRA 07]. The population increased in America first through the ports in successive waves. This dynamic occurred early in the Northeast Region of Brazil with the Portuguese (and later the Dutch) and, in a more gradual manner, in North America. The British colonies opened up through their ports, from Boston to Charleston at least; and it is with the uprising of the merchant middle-class in the ports, that the

United States of America was born. If the population in the American megacity is so high, it is due to the waves of emigrants originating from all over Europe but, with a closer look, the ports are the first to fill up, then the population is redistributed within the cities, however, this dynamic does not explain the dense occupation that we witness along the coastline. The American growth in population in this case is due to the tourism development in the beach resorts that followed the classical hierarchy over time. Nantucket, the fishing village of cod and whale (Moby Dick), is very different from Atlantic City in New Jersey, which, in spite of its size, is up to the evocative names given to its resorts such as Seaside, or Carolina Beach, etc. Elsewhere in the colonial world, the same logic applies, from Indian trading posts to the Cape region or the Chilean coast, not to mention Shanghai or Singapore. Let us add to this that the overturn of the Chinese political direction had a significant influence on coastal regions by increasing their importance thereby favoring the growth of large ports. This phenomenon is a recent one, yet it fits within the logic of contemporary globalization, which involves the opening up of the seas and coastlines. Furthermore, it reinforces the poles while the growth of tourism, since the end of the 19th Century, progressively filled the gaps and increased the pressure imposed on coastal zones. France, along with the Spanish coastline, is particularly exposed to this phenomenon. If we look at it in numbers, in 2006, the coastal urban communities welcomed 10% of the population, i.e. 2.5 times more than the national average; and if we include overseas territories (excluding French Guiana), the numbers amount to 317 inhabitants per km^2. This highlights how strong the pressure on islands can be. If we distinguish the coastal fronts, it is easy to see that the pressure is more acute along Mediterranean coasts than on the Atlantic front. The magnet effect of coasts combined with the attraction to sunny climates suggests that this increase is strongly related to the expansion of tourist activities. The growing attraction to coastlines is not without consequences for the resource uses both in terms of space and ecosystems. Given the rarity of the space and the fragility of these ecosystems, the competition between users generates conflicts that ought to be managed intelligently. This involves respecting each of the user groups as none could be deemed illegitimate.

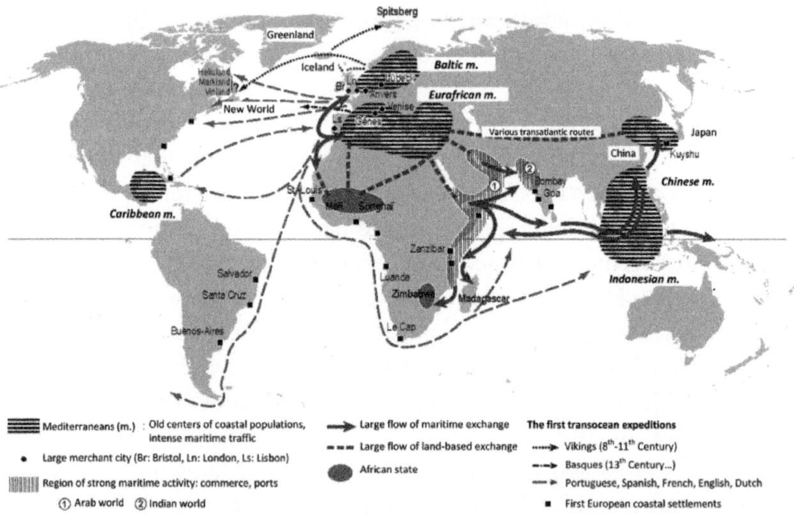

Figure 5.7(a). *Population increase along the coastlines at the beginning of the 16th Century (according to J.P. Corlay)*

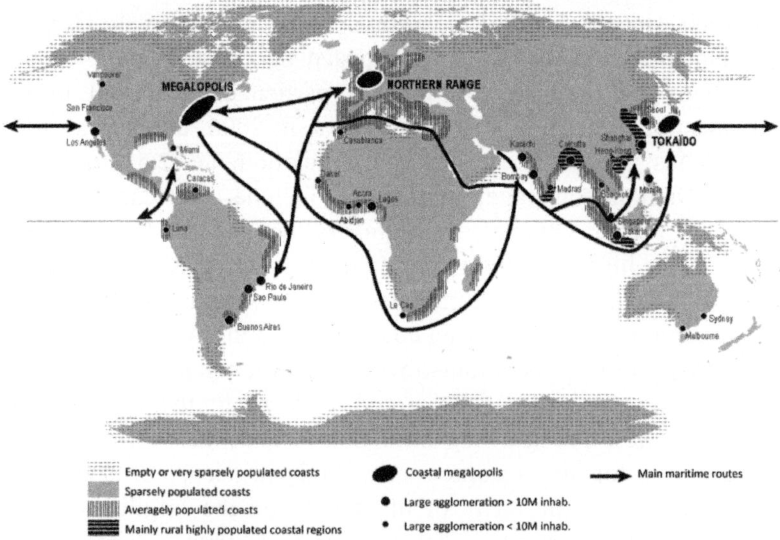

Figure 5.7(b). *Population increase along the coasts at the end of the 20th Century (according to J.P. Corlay)*

5.3.2. *From resource use to competing human activities*

The reasons behind conflicts in resource use are varied but they are intensified by the nature of the coast itself, as a coveted narrow and fragile strip. After all, everything crystallizes from one end to the other of the shoreline, on the shore facing the ocean but also in the sea itself. This consideration forms the baseline of the concept of coastal zone. On the shore, what tourists seek is the proximity to the sea, which is used as a slogan to sell apartments in Merlin, in Vendée, "sea view and feet in the water". In the sea, and along the coastal zone, the activities range from strictly professional (fishing) to various leisure activities, such as windsurfing with the more modern varieties of surfing, surf casting and other forms of leisure fishing, and of course water skiing, which forms a demanding and powerful industry. Tapping into these activities reveals many conflicts, and it becomes clear that these are more or less difficult to solve (or at least to manage) with a bit of good will and interest in reaching a consensus. These are human issues, first in the heat of the conflict, but they also regard collectives, municipalities and the State in the case of France, or local "communities" and federal States including coastal States in the United States of America.

5.3.3. *The modernization of fishing and its effects on the oceans*

Although fishing is a traditional activity along the coasts and at the land–sea interface due to the potential richness of coastal waters, it is also an activity that has encouraged globalization, with fleets tracking shoals of fish across seas all around the world. This industry, therefore, encourages the activities on the seas both intensively and seasonally. Finally, it is often a controversial activity due to the abuses it can embody. So, fishing activities today are inseparable from the protective measures and policies that accompany them to ensure their sustainability. Fishermen are first of all predators; it is, therefore, not surprising that they sometimes spark criticisms from people who do not know much about this difficult profession, in which fish tracking has become all the more difficult as technical developments started accumulating. However, this only shows one side of the argument on the sustainability of the harvesting of the ocean's living resources. There are many conflicts, and these are often emphasized by the media, in which case it is debatable whether the real target is actually the one being reached or not. When Greenpeace blocks the route (and the catch) of Japanese whalers, it is easy to forget that the development of small-scale fishing for many "poor" fishing communities plays a similar role against the resource in question. However, other measures can be used to better control the resources and their use (Couliou in [MIO 12]).

5.3.4. *The rise in coastal tourism and its effects on coastlines*

More recently, the tourism industry is having a high impact on the coasts, further into the coastal water where nautical activities are found in all their forms, with cruise ships being the extreme embodiment of this development brought about by the very quick economic growth of the last decades. This is not without consequences. Various forms of pollution are produced for a start, which affect the quality of the water and thereby gives rise to conflicts of uses with on the side of the polluters, the tourism sphere [BUH 09] and the farmers discharging chemicals in the catchments upstream. A representative example is that of the pollution of beaches by green algae, especially in the North Finistere (Figure 5.8). But, even more controversial is the growing occupation of the shoreline by individual houses, more or less regular estates, more or less recent beach resorts, as we see them along the beaches of the Mediterranean (Spain, the Languedoc Coast, the Italian Adriatic Coast, the Croatian Coast, etc.). These developments represent an unoriginal way of unifying often sparsely populated environment, except for fishing or mussel-oyster growers, and which, as can be expected, benefits from little economic growth, due to spatial competition on the foreshore and due to the water quality most importantly.

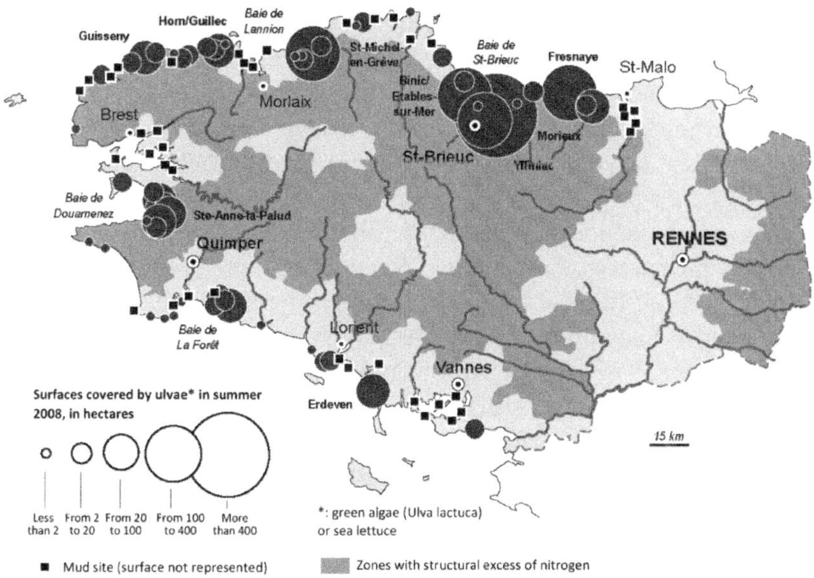

Figure 5.8. *Green algae (Ulva lactuca) in Brittany, summer 2008. This type of pollution is representative of the emergence of a form of agriculture that is closely integrated within the industrial world (phosphate fertilizers and animal feed sometimes obtained from industrial fishing). The different volumes indicate the dissymmetry within Brittany with its short and steep rivers flowing into the Channel. The algal bloom in spring and summer is occasioned by a disruption of the balance. This constitutes a threat for biodiversity but is also difficult to resolve without affecting social dynamics*

5.3.5. *The sediment deficit and the growing demand for sediments with its effects on coastlines and the foreshore*

The need for marine aggregates, silicates and calcareous sand (or corals) in building industry, roads and for the refilling of beaches is growing. This only gives rise to a few conflicts, but it presents itself as a challenge for the future. In a similar way as States basing their economy on oil and mining explore the continental margins, the sandy resources are coveted by national authorities and private companies that are aware of the difficulty in obtaining these resources. The sources of sediments are first of all the continental shelves but as these have been widely used and the regulations around extraction are as numerous as the needs for sediments, we have naturally turned toward the most abundant source of sediments in principle – and for the philistines – the source offered by the sea. Sediment extraction is not a recently discovered activity: for decades, part of the coastal urbanization, in particular in the case of many beach resorts, is the result of this practice. It was convenient to dig into the bordering dunes (many caoudeyres are nothing more than quarries), or even the beaches themselves. For instance, when it was attempted to refill the beach of Baule (almost for the first time in France at such a scale), it was decided that approximately the same amount than what was taken in the years 1950 and 1960 would be deposited. Similarly, the improvement of acidic farmland in Brittany was achieved with calcareous sand extracted from the bottom of the sea in shallow waters, named maerl; many small Breton estuaries still undergo this movement on a periodic basis and, in theory, in a controlled environment. The monitoring of sediment extraction is the double consequence of the particular care borne to the environment over the last three decades and to the realization that the sediment scarcity on the seaside is not a scientific fantasy: the growing effects of coastal erosion are the expression of it [BIR 85, PAS 81, PAS 93]. Coastal erosion not only occurs locally but at various scales and is the observed consequence of a given depletion of sediments, which itself is related to the combination of man-made structures around streams (dams, spikes, dikes which increase the energy of the water flow, etc.), of the control exerted over watersheds (plantations in some Moroccan wadi, for example, which have slowed down the sediments on their way to the mouths of the wadi and thereby contributed to the considerable erosion of the large beach of Agadir), but also to a certain degree of depletion of the shallow pools offshore, as if the end of the Flandrian transgression had exhausted the last stocks available. However, this remains a debate among specialists.

All these elements contribute in explaining the gravity of the problem of sediment extraction offshore. The needs are considerable; the European Union (EU) recently reported that no less than 50 million m^3 were extracted each year from the North and the Baltic Seas at less than 60 m depth, and these needs are likely to increase over time at least as a result of the demands on the coasts [EMS 10]. All the large companies, e.g. Lafarge in Europe, are looking for areas from which to extract sediments. These are found under two categories, either as fossil sand and gravel deposits left to settle over the latest transgression (or over repeated transgression-regression oscillations during the Quaternary), at depths that are known due to increasingly precise cartography; or as "living" structures, i.e. all the shoals and "ridins" of the platform mobilized by the swells.

This type of exploration tends to increase the cost of sand or gravel by cubic meter, and is therefore in conflict with the technical capacities for extracting and transporting sediments to the coasts. It is also in conflict with the constraints from the regulations for environmental protection due to the benthic life that inhabits it, to the reluctance of fishermen themselves and finally to the mobilization of associations for the protection of the environment.

A typical example of this is the extraction site targeted by Lafarge at about 40 m depth off Groix, facing the Plouharnel beaches. In this area, the fear is to see the shoreline retreat through adaptation of the foreshore profile to the created trench. It was a similar worry that mobilized in vain organizations opposing the exploitation of the ridge of Pilier off the Island of Noirmoutier.

In the future, the exploration of potential sites will have to be carried out in a more flexible framework offered by the integrated management of coastal zones. It represents, along with offshore wind farm sites, one of the greatest challenges for marine spatial planning, which is still in its infancy. It is also a way of improving our understanding of the foreshore dynamics. However, in the United States or the Netherlands, for example, where sediments are used regularly to refill the beaches as a common practice to fight the sea, the volumes needed are so massive that the requirements there again are likely to increase, especially in the face of sea level rise. A slight increase of the level of the ocean could accelerate a worrying phenomenon.

5.3.6. *The globalization of trade and its effects on coastlines*

Today, more than a billion tonnes of goods get shipped around the oceans and loaded onto the various ports and harbors. This realization arises from a double dynamic: the first that pushes toward an increasingly extreme integration of the whole world on the economic level; and the second that was initiated at the end of World War II and saw the capacity to load ships increase dramatically at the same time as old freighters were giving way to increasingly more specialized boats (from ore and oil tankers, all the way to the very modern container ships). Globalization is both a consequence of a general loss of toll rights, and of the deregulation of financial market in the context of the global expansion of capitalism; it also marks the accelerated integration of the economies at the regional level (Mercosur, EU); the appearance of new competitors and new markets in Third World countries, emerging markets that influence the global economy (China, Brazil, Russia, India as well as Indonesia and even Nigeria). Finally, it is a response to the globalization of production systems for the large (globalized) multinational companies in that these are more flexible organizationally and spatially. So at the beginning of the 21st Century, our economy traveled by sea, carried around by all these technical and technological revolutions with the shipping container as their flagship. Our economy travels by sea as this is the most economical means of mass transport and that for which the costs have gone down most dramatically. American engineer Malcolm McLean and his team in the 1950s probably had only a vague idea of the extent to which the little transport-all box they were devising would impact the world. A technological revolution is underway as the various actors of the maritime transport, i.e. ship-owners of regular routes, forwarders and handlers, reorganize their working practices around this technical innovation. The ships that used to transport these containers at the end of the 1960s could not fit more than one thousand boxes, whereas large containers today can fit up to 18,000 units. Needless to stress the demands in terms of the size of docks, platforms and intermodal transportation infrastructure to fit 18,000 trucks on land. The stake for the ports becomes considerable. The territorial animation is at the scale of hinterlands: these ports define the grids, and their influence is significant as they combine transportation systems and are indirectly responsible for the level of activity inland. These ports require space and on this level, their appetites meet others' and from there, conflicts of usage germinate that are sometimes difficult to solve. They mostly relate to the opportunities for spatial expansion where stakes are high and where competition and the balance of power form part of the game played by the lobbying groups.

Two examples in France can illustrate this model. First, the port of Le Havre with its great (ongoing) Port 2000 project was able to claim the last mudflats of the estuary to improve the conditions of its service, in particular for container ships. This occurred as the supporters of the port were organized more efficiently than the opponents, as these were scattered and therefore less efficient. The latter comprised more than just organizations for the protection of the environment. The second example refers to the debate around the expansion of the port of Montoir into the mudflats and marine grassland of Donges-Est that seemed to be never-ending but was concluded by the victory of the opponents to the expansion [DES 09, MIO 98]. Of course, we are comparing a port whose traffic approaches 100 Mt with one which barely reaches 30 Mt. Moreover, the port of Le Havre is also situated on the route of the Northern Range of the Channel-North Sea ports that opens onto the traffic and economic development in the heart of Europe (from Le Havre to Hamburg) and, therefore, the stakes may appear quite different. However, the conflict arose around the survival of the swamps and mudflats, around their long-term presence as the basis of various trophic chains since they harbor nurseries that are indispensable to a "sustainable" fishing industry to mention only one example of amenity offered by this habitat.

Contemporary ports can only expand to the detriment of wetlands, which have very high stakes associated with them. Indeed, these environments are increasingly rare and affected by all the different types of pollution emerging from watersheds. Their conservation is considered as absolutely vital in the perspective of sustainable development especially as this concept increasingly underlies policies. Ports can alternatively expand by gaining territories at sea, in the form of intensive accretions; however, these are not without consequences for the quality of the marine environment. Sediment inputs also require dredging at sea in order to gain access to the port following the depths used for docks, as ships need to load and unload goods but also attempt staying motionless for as short time as possible. These needs underlie the progressive decrease in floodplains in the Bay of Tokyo, which offers the best example of the gradual silting of a bay. The same phenomenon can be observed in a different context in Singapore, the first port in the world, or at least in the Chinese economy, that is largely driven by import-export; this port requires considerable surface input. With the help of the particular political regime in place in China (and in spite of its theoretical engagement in the integrated management of coastal zones), the preservation of the environment is apparently not a priority. This is in contrast to Europe, and France in particular, where environmental protection influences the way development policies are deployed. Uses and conflict of uses, beyond the local environments in which they take place, are nothing more but the expression of the balance of power within societies. Finally, the issue is cultural and clashes with the idea of globalization as a phenomenon that reduces cultural differences.

5.3.7. *... as a transition*

The increased attractiveness of the coastlines and the race of economic imperatives offshore have strongly unified seas and coasts in the expectations of people and contemporary societies. The conflicts in uses are merely a reflection of frustrated expectations: some think that economic progress ensures a future, whereas others go as far as envisaging economic downsizing to ensure a sustainable future for all. In the latter's opinion, it is the survival of the planet that is at stake. It is as if the millennialism, which is often evoked by historians about the 11th Century, has risen from its ashes: our new millennialism is referred to as global warming for the optimists and climate change for the pessimists.

It is not possible to understand our contemporary heritage policies without taking these concerns into account. These are often more latent rather than spelled out [GUN 09]. For some, the solution lies in the fencing off of flagship ecosystems and habitats that need to be preserved at all costs; whereas for others, who are indifferent to these arguments and are carried away by the global dynamic of "development", growth brings economic and social well-being. In between the two opposite positions, we find various levels of endeavors which often converge in such a way that they intend to secure a future while preserving the past. Underlying it all, we find the whole debate around biodiversity a very topical emerging issue [GOU 10, GOU 12, TOU 12].

To a certain extent, this is the heritage vision that will condition the policies we will now discuss. These policies cannot remain indifferent to the current changes observed; however, they are not directly linked to them. At least, this can be argued in a few lines. The fact that the debate is temporarily appeased does not stop us from feeling its effects on the ground.

The Earth's climate has been warming undeniably since the last third of the 19th Century: the changes in climate show at least an intermittent trend with a more pronounced warming of the last two decades. For a change in climate to occur, real structural changes would be required in particular in the large pivots of the climate system, such as subtropical anticyclones and, in the ocean, the broad cold and warm currents. Currently, nothing shows that this is underway, not even climate models. Incidentally, this is a debate whose scope and relevance are political and for which scientists should not be dragged to a place which is not theirs. In terms of coastal issues, the main relevant element is sea level rise that remains a slow process. The sea level started to rise in the 1880, as shown by tide gauges; however, these are few and are all situated in the great ports of the Northern hemisphere. The accumulation of tide gauges confirmed the trend of sea level rise that is now monitored by satellites. The last IPCC predictions remain moderate: 50 cm at most by the end of this century. Of course, the surfaces situated at the lowest altitudes will gradually be submerged. We know which are the places concerned and we also know of a variety of solutions to address this problem. The remedies are partly technical in terms of resistance against the sea, and partly political, in the adaptation by retreating away from those submerged areas. None of these elements come into conflict with the heritage policies we are about to explain; on the contrary, these still need to be integrated within the development process.

5.4. Management models for the protection of coastlines

The situation described above undoubtedly appears very bleak. It cannot be otherwise given all the types of pressure exerted on these environments, with consequences that require attention. The concerns were for a long time rather focused on heritage, proper to a western society dominated by elites of birth or fortune. The emphasis was on the heritage values of landscapes, for instance, of sites, monuments without going beyond and recognizing the shorelines and oceans for what they were, in their specific character. From the 1960s, this perspective changed as the attention was borne onto sites that should be conserved for their high heritage value but, more broadly, it was the natural environment that became the common human heritage. The shorelines and seas are also part of this new logic [MIO 93, MIO 09].

5.4.1. *The stages of the realization*

The realization was slow and progressed, as the various stages and types of degradation were noted, with incidentally, a growing trend, in some places, to exaggerating the effects of the process. With the rise of coastal tourism, local organizations of residents started reporting, not without some ambiguity, the high pressure on the coasts and the degradation effects that have often resulted from it. The slow continuous erosion of the shoreline was often revealed after the local residents who were suddenly confronted with the sea, not just to admire it, rose together to demand protection. This realization essentially marked the brutal overturn of a farming society that was indifferent to the sea, except in the case of dikes for the protection of marshes and polders, toward an urban society, with their characteristic hedonism, which would welcome the image of stormy weather on the winter coasts as long as the villas are not damaged. Few went as far as thinking about the implications of this. This regarded the local scale: the gestation period for the need to take environmental health into account was very long. It is in fact from the top that the movement was initiated, as much in the United States as in France as they both became implicated in the 1970s. The paths followed by these two countries were different in the short term, but ended up converging. Both aim to preserve the quality of the environment and the coastal environment without constraining the means to economic development too much. Of course, the conservationist ambition was deeply rooted if we look at the secular history of the British National Trust. However, its purpose was more anchored in the conservation of the heritage (of monuments) rather than of nature itself, and it was a long time before the creation of the Heritage Coast included coastal sites and enhanced them for the public.

So, coastal and marine protection is a new construction, and it is still difficult to dissociate it from oceanic coasts, both for political reasons (the position of the States toward their coastal zone) and cultural reasons (the populations living on the coasts form a minority, whatever statistics and maps say: we can live 10 km away from the coastline without ever looking at it; the "attraction" of the coast often needs redefining, especially in France).

5.4.2. *The implementation of the Coastal Zone Management Act in the United States*

It was in the United States of America, at the end of the 1960s, that determining steps were taken for the protection of coastlines. The Stratton report, *Our Nation and the Sea*, was commissioned by President Richard Nixon and defined the framework within which the various partners involved progressively built a policy both for the protection and development of coastal zones. The *Coastal Zone Management Act* was approved by Congress (the Senate and Chamber of Representatives) and became federal law on October 27 1972 when Nixon ratified it. The law in itself appeared to have been imposed through a top-down approach and to be contrary to the coastal States' interests, as these were always very proud of their autonomy. In practice, everything turned out differently.

The American text is innovative on many fronts: it solemnly states that the coastal zone is resource rich, from the point of view of natural, commercial, recreational, ecological, industrial and landscape resources, and at the same time subject to growing pressures with measurable effects (depletion of marine biological resources, constant and "regrettable" ecosystem change, of the spaces accessible to the public, coastal erosion, etc.), ecologically fragile and therefore very sensitive to human pressure. In summary, it is a heritage that is crucial to the survival of the Nation, and should be managed with rigor. The *Coastal Zone Management Act* was supposed to fill a gap left by the absence of regulation framework to solve coastal problems and to encourage the coastal States to exercise their full authority on the environments and habitats forming the coastline. On its own, it is an "open" federal text that encourages local initiatives. In practice and in the long run, the *Act* is a real success story that reinforces protection without halting development. Furthermore, it is ground-breaking in that it recognizes the singularity of the environment and, on the political front, the authority of the States.

About 20 years were needed for every single coastal State to adhere to the *Act*, not without some level of reluctance but, sometimes, with the intelligence and pragmatism necessary to understand what was at stake. Adherence to the *Act* brought federal aid that provided support to local initiatives, in an effective balance between central authority and the structures put in place in each of the States. At the center, collaborative practices, all based on the lobbying allowed by the American law between the *Office of Ocean and Coastal Resource Management* (today known as *National Oceanographic and Atmospheric Administration* – NOAA), where politicians appointed by the President and officers convinced of the importance of their mission, such as organizations for the protection of the environment, professional groups (*National Association of Homebuilders*, *US Corps of Engineers*, *American Petroleum Institute*, etc.), sections of the Congress responsible for coastal issues and of course coastal States, some of which were pioneers in implementing strong protectionist measures (the State of Washington, California and Maryland), all work together.

5.4.3. *The slow development of a centralized concept of coastal planning*

France shifted, in a period of about 40 years, from a state-controlled and centralized concept of what we term "coastal planning" toward an allegedly (but not truly) decentralized policy of coastal management [MER 09, MIO 99]. Beyond the ambiguity of the vocabulary used, these are actually two different approaches to territorial planning. The first, traditionally Anglo-Saxon and in particular American, advocates that within a federal State, coastal States should be able to legitimately implement their own policies, which are *de facto* "decentralized"; and on the other hand, it preconizes an approach relying on the long-established national construction and on true mistrust toward its techno-bureaucratic elites when it comes to local practice, and which considers that steering and impulsion should come from the top. It is the French "top-down" against the American "bottom-up" approaches. In reality, things are not as black and white as described and policies end up converging with the overarching aim of a more sustainable management of the environment and of marine and coastal resources.

Coastal planning is inseparable from territorial planning in France, of which it is only a marginal part. It encompasses the assertion of power of the State after World War II, and more especially after 1958 when DATAR (French Department for Territorial Planning and Regional Action) was created, this devolution of territorial planning to the regions, whose numerous title changes in the last decade say more than any discourse about the uncertainty that revolves around planning. Starting with the State, it therefore often embodies a royal vision, steered from the top and, to a certain extent, indifferent to regional cultures and to the concerns of local populations. Its aim is to develop economically by choosing functions that are likely to have a ripple effect that could nevertheless be controlled.

Coastal management of the coast of Languedoc-Roussillon is the best example of such a process. It was carried out as part of an interministerial mission, known as the "Racine Mission" after the name of the high officer who was in charge of the mission (Decree of June 18 1963). The mission had several objectives. First of all, it was to structure the coastline of Languedoc in order to attract tourists, especially foreign tourists (northern Europeans), who normally spend their summer on the Spanish Costa Brava nearby. This influx of tourists is a source of income. However, the choice of tourism developed to the large coastal scale between the Alberes mountains and the Rhone delta was supposed to generate jobs, both directly while building the beach resorts, and indirectly through the jobs produced by the demands from the tourism industry. In terms of planning, the operation was to take place in a practically virgin field (at least considered as such by many who do not care much about the traditional cottages thriving in popular environments). Therefore, the tourism units are easy to design and draw on paper, with planners and architects deciding how tourism-based urbanization will unfold. People are not allowed to develop in the direction they want as following the typical liberal "laissez-faire" of the coasts. It goes without saying that the perverse effects of "land speculation" are largely mitigated within the resorts and should not be extended to the unspoilt margins that will function as breaks from urbanization, and will constitute "windows" in which nature is protected and offered to holidaymakers. Finally, the aim was to build tourist units around the leisure ports, which there again creates opportunities for the nautical industry and generates jobs that do not rely solely on the tourist season. We are faced with a proper coastal planning operation, if we use Paul Claval's definition of planning: "the set of concerted measures which regulate the use of space and its planning so as to ensure the full thriving of individuals, to make social life easier while minimizing frictions that might arise from the distance or the proximity of antinomian activities and to avoid disturbing the natural balance whose destruction would be, in the short or long term detrimental to the collective". After all, planning is a deeply proactive, political act that mobilizes large amounts of space, humans and capitals and betrays an interest in "managing" space that could not be left for the set of stakeholders to dispose freely of.

This vision of territorial planning persists in France in the texts and in the mentalities of people, at least at the State level for several decades; however, the golden age of the 1960s is over. The interministerial coastal planning mission for the Aquitaine coast did not achieve the same success in the next decade. This was due to the environment that was undoubtedly different and more difficult to plan, but also to the attraction, as the Aquitaine coast does not offer the same climate potential and is not situated in a major transit toward the Iberian Peninsula. This decade saw the launching of a slow dissolution of the concept of management in the French way. Indeed, the Piquard report provides a key to understand this phenomenon; its title "Shorelines, planning perspectives" [PIQ 74], however, its content, very balanced, expresses concern around the excessive intensification involved by certain activities, especially tourism. This report opens up the opportunity for the "natural third-party" on the shoreline and announces the creation of a Coastal Protection Agency (*Conservatoire du Littoral*).

Planning, as conceived during the glorious 1930s, is over; the experiences following this which form part of this positivist and even productivist logic more or less failed (for instance, the Models for the Development of the Sea – *Schémas de Mise en Valeur de la Mer*, SMVM), the Coastal Law of 1986, cornerstone of a policy that remained stuck in a "planning" logic (as seen from the title) brings a change that slowly and reluctantly, due to the nature of the State itself in France, toward a new vision of the shoreline, i.e. integrated coastal management. Management comes from the English term: the ambiguity of the language also expresses a "culture", which in France is very centralizing.

5.4.4. *The emergence of a policy of integrated coastal management, from the global down to the local levels (and vice versa)*

5.4.4.1. *Agenda 21*

It was after the World Summit of Rio de Janeiro in 1992 that the need for a global approach to marine and coastal issues was felt. Each and everyone was then able to denounce the very heavy pressures on natural environments, as was already stated in the Stratton Report and as was confirmed by the CZMA in the United States. The pressure on the habitat (fragile and rare), here the coastal zone refers to both sides of the coastline, pressure on the environment, both terrestrial (marshes and coastal dunes) and marine, and by extension on the sea far offshore due to the fish resources. Quietly, but with the more aggressive and denouncing support of non-governmental organizations (NGOs) dedicated solely to the protection of the environment, the participants in the Rio Summit attempted to conciliate the points of view of the stakeholders according to a logic which is, once again, borrowed from the United States, but behind it is the broad Anglo-Saxon management form for a "responsible use" of resources, both physical and biological. This gave the signal to the integrated coastal management.

Agenda 21 is the reference document of the World Summit that was held in Rio de Janeiro. It constitutes in a way the guidelines of the international community for the 21st Century, a "global strategy for sustainable development" that should be articulated from the United Nations down to the smallest coastal community, but also upward from the latter, as their practices need to integrate within the pre-established framework that requires the acknowledgment of economic, cultural and social local realities. It is the two-level logic with the experience from the bottom (bottom-up) and the recommendations from the top (top-down). In practice, Agenda 21 also prematurely ratified, right after the fall of the Berlin Wall put in motion fundamental changes, intensified globalization. The ambition is global, it concerns the whole planet and touches on all the problems it will experience and will have to face. For all these objectives, the only realistic and relatively empirical method is international cooperation. The latter is the only option likely to guarantee a certain level of security for the planet as well as a certain socio-political balance. There is no mention of extending democracy to the whole world, but the greater transparency needed in the process is mentioned, which is a way of envisaging the diffusion of democratic principles. Sustainable development is the overarching aim, and it requires the more intensive economic development of developing countries, so as to fight poverty and make health policies more effective, not to mention (but it is less clear) the issue of demography. As we can see, the ambitions of Agenda 21 are broad and inclusive rather than totalizing, but are largely inspired by Anglo-Saxon cultural practices. Everything that touches on the environment is considered with increased interest, starting with concerns around climate change (or global warming). When we refer to "global change", we enter the domain of relations between humans and their environment. What policies should be used to avoid accelerated deforestation, what principles should be accepted to avoid carbon emissions in the atmosphere, and how to face sea level rise?

This is a real revolution at the planetary scale with the breaking down of Agenda 21 into a series of local agendas aiming to set the local conditions for sustainable development. Each regional unit is required to address these (Europe is not exempt from them) as well as each State. The French "*Grenelles*" echo, albeit from afar, chapter-by-chapter, each topic addressed. Chapter 17 in Agenda 21 is dedicated to the oceans [CIC 98, MIO 12].

5.4.4.2. *Integrated coastal zone management*

Integrated coastal zone management is an emerging concept that tends to consider that the coastline can only be managed if it is not seen as a system in which all the elements are inter-related. The idea originates in the Anglo-Saxon culture and is inspired, in the long run, by the principles set in 1972 with the enactment of the *Coastal Zone Management Act.* This idea also punctuates the next two decades and ends up, during the 1992 World Summit in Rio de Janeiro, with the insertion of the integrated coastal zone management in Chapter 17 of Agenda 21. The emergence of an integrated management approach is closely related to the growing awareness of environmental problems. Indeed, it is with the vision of managing all the marine and coastal resources responsibly, and with the vision of sustainable development, that the interest of the collective was seen as more important than individual interests. This logic follows the principle that exchanges and dialogues between stakeholders in the coastal game were keys to good governance.

Integrated management of coastal zones relies on a vision of governance that aims to draw on bottom-up initiatives, while leaving the motivational role to the top, especially in terms of financial incentives. The concept of integrated coastal zone management was designed first at an international scale under the auspices of organizations depending on the United Nations, then taken over on the ground, especially in the South, by NGOs. It is subject to experimental programs at the European scale and at the scale of most of the member

States of the EU. France joined the experiment relatively late, as its centralizing traditions were slowing down local initiatives. However, after a decade, and since the proposal made in 1995 in the Bonnot report ("For a global and coherent coastal policy in France"), the principle has progressively made its way to the concensus, due to a decentralization process, local initiatives created fertile ground for a new vision of coastal territorial management, seen both as terrestrial and marine. Today, it appears that the integrated management of coastal zones, if implemented with common sense and flexibility, could bring solutions to problems that the "models for the development of the sea" were not able to solve. So, the rise of principles based on an integrated management of coastal zone approach should be seen as the expression of the progressive convergence of visions under the pressure of globalization that rightly supposes all global problems be tackled globally. Integrated management does not disregard global warming, nor sea level rise, with the risks associated with it. However, it goes further in demanding tolerance for practices that mostly reflect local cultures that should be respected. In this respect, integrated coastal zone management rather expresses the triumph of a decentralized vision of management, and if France adopts it as well, it is both out of frustration (as none of the solutions that were steered by the State produced good results) and also in the name of political realism that acknowledges the construction of Europe, the principles of local initiatives, that are more easily accepted in States where provinces have historically had a strong level of autonomy.

Integration is a basic principle that is sectorial, administrative, spatial, international and that should indicate the strong presence of scientific research. In this, it links together research and sustainable development whose integrated management of coastal zones is one of the most prominent aspects. It is sectorial in that the participants are selected on the basis of their being organized enough to debate and influence the decision that will unavoidably be political. The transition needs to be made from a logic of poorly solved conflicts (or conflicts considered as impossible to solve) to a logic of transparency, dialogue and reciprocal concessions. The norm is to look for a compromise; the logic is no longer to "lobby" in the negative sense that is too often attributed to this term. It supposes a world of organized responsible fishing, able to understand the stakes and to help the participants to accept decisions that are sometimes difficult; this is true for all professions (the world of ship-owners and ports, of real estate, etc.). It is administrative as too many sea-related administration bodies (and the case of France is crystal clear) hold diverging positions that neglect the viewpoint of the professions of the sector they administrate. The 17 administrations dealing with marine and coastal issues in France have to discuss constantly in order for the higher interest to win over small groups. A General Secretariat of the sea, with the guidance of the National Council of Coastlines, could have a very positive impact if its local delegates were responsible, well informed and endowed with a spirit of innovation. On the spatial scale, integration should enable considering the totality of the space qualified as coherent, linking the land to the sea and detaching its action priorities toward a sustainable form of development, i.e. respectful of the ecological balance responsible for the given resource, limiting exploitation (of biological species as well as of the space itself). Spatial integration means a geographical area of dialogue but not necessarily, as many people assume, spatial "zonation" or the reinforcement of the constraints on the limits which at sea are difficult to determine. Finally, it involves the integration of science in the stakeholder discussion, and of course integration within the scientific community, as the latter still too often reflects paralyzing divisions. Natural and life sciences should have an open dialogue with human and social sciences in order to go beyond pure science, to propose solutions to all non-specialists who are sensible and intelligible to each and everyone involved. This is not the smallest of stakes in the integrated coastal management; the latter is an iterative and highly democratic concept in that it implies that each stakeholder should be the master of its destiny. This new form of governance is only a tool when considered in isolation, as it is broad and decentralized. It enables both economic development and the protection/conservation of the shorelines and seas as heritage in the more common sense of the term, i.e. what humans inherit and should conserve and strive to improve. It is a simple and realistic tool, not remotely similar to the gas factory that some administrations, the French one in particular, put in place and which perpetuates the reasons that brought criticism to their outdated practices.

5.4.4.3. *A toolbox for the protection of seas and coasts*

Behind the broad criticisms highlighted above, an arsenal of various measures taking into account the heritage value of places and ecosystems can help in constantly monitoring the changes (or even the "development") to define thresholds beyond which a given space, environment or a species is threatened. These measures are derived not only from political decisions at various scales, from the local to the global, for instance, with the creation of local urban planning (LUP), but also depend on territorial coherence in France (SCoT) with a vast area such as the Mediterranean that addressed by the Blue Plan, which was itself derived from the Barcelona Convention in 1975, all under the care (although that is not always the case) of the United Nations Environment Program (UNEP). These measures help not only States, but also increasingly NGOs dedicated to the protection of the environment, in their protective measures as States often nourish preoccupations that are more turned toward development and that have to be adapted to the aggressive attitude of NGOs as best as possible. Everything cannot be described, knowing that everything is subject to protection, from shorelines to coral reefs, from mangroves to maritime marshes: some "tools" can, however, illustrate the action, such as in France.

5.4.4.3.1. The laws for the management of coastlines, a French model of management

Law Nb 86-2 "relates to the planning, protection and enhancement of the shorelines". It was ratified on January 3 and sealed a series of texts regarding the regulation of the uses on the seaside. It was very controversial in that it constitutes, first and foremost, a compromise between those who wanted to "plan" the coastline better and those who, on the contrary, wanted to "protect" it better. The law was reinforcing a very "regulatory" vision of the shoreline following the line of the Order of the Navy in 1981 which stated that the coast was an integrant part of the king's territory and therefore, at the time, of the State as well. This is not a law aimed at the particular concerns we have been discussing of rising awareness around the growing pressure imposed on the French coastline; it is a text that rightfully intends to set a number of limits not to be surpassed in terms of the occupation of the coastal space. This explains the reluctance observed toward the law, as it was issued at the same time as decentralization entered an active phase and the State was accused of taking over where the locally elected were determined to make use of their new freedom.

The Coastal law is relatively short, with three titles and 42 articles through which the philosophy transpires. The law cannot be mistaken as guidelines for the global management of coastlines, as this will not be issued until the beginning of the 19th Century with the implementation in France of the integrated coastal zone management, and the law on which it is based remains a reference point.

The law is very specific and similar to the law regarding mountains voted by Parliament in January 1985. The first article identifies in the coastlines "a geographical entity" that "requires specific policies [...] of public interest". In addition to the particularity of coastal areas, the objective is to manage it so that it is accessible to all the members of the public and so that (although it does not appear clearly in the text) it is considered as part of the National Heritage.

Apart from the order of the principles set, the law serves planning, protection and enhancement purposes for coastlines and each reader interpreted what he/she wanted from it. For example, the local authorities read in it, at first, and similarly to most people, that planning referred to the pursuit of coastal planning as a way of perpetuating the development of tourist activities, albeit in a slightly more regulated environment, which was probably perceived to be too regulated in their eyes. The law can broadly be subdivided into two parts. In the first part, the conditions of the "development" are specified; whereas in the second part, the principles of effective protection of the "characteristic" coastal environments and ecosystems are highlighted but without specifying the limits to which the effort will be carried out.

Title 1 is related to the planning and protection of coastlines. It is the main and only title we will be discussing here. It adapts "some provisions of the Planning Code", i.e. the main objective of the law.

How could the conditions of economic development – and in particular tourism – be maintained over time (as the law is supposed to refer to sustainable development)? Some of the indications given provide a better understanding of what can be done. First of all, the claim that there is a certain "capacity" in urban or urbanizing areas: this capacity is not precisely determined, but it implies that natural areas should be preserved and that areas where traditional activities dominate (agriculture, forestry, etc.) should be protected. It is a way of freezing these areas in time and of alleviating the pressure. Therefore, it is claimed that breaks from urbanization are nowadays necessary: the extension of buildings can only be carried out in such a way that it fits with what is already there, as a form of "new cottages in harmony with their environment". It is also specified that in areas close to the shore, extensions can only be built if they are used for economic activities that require a close proximity to the water. Finally, outside of cities, constructions are illegal on a coastal strip of 100 m. All these elements do not aim to put a stop to urbanization, but rather to channel it in each and everyone's interest: it is no longer allowed to build anything anywhere, which does not mean that any development is prohibited. In parallel, the significance of these areas and environments is emphasized; the logic is a consequence of the principles underlying the Piquard Report in 1972 and the creation of the Coastal Protection Agency ("*Conservatoire du Littoral*") in 1975. Article L 146-6 was controversial and offended many local officials. Indeed, it relates to terrestrial and marine areas and to sites and landscapes with remarkable features or characteristic of the natural and cultural coastal heritage as well as ecosystems necessary to biological balance. These remarkable and characteristic areas should be spared from any form of serious occupation, and all the planning documents should take them into account as soon as these areas are described and spatially defined. As always with laws, decrees containing many articles (especially of L 146-6) were delayed and caused many issues; in this, it appears that the law should not be imposed as a unit but should rather constitute a framework for thinking about issues, for debating, and finally, for acting.

To a geographer, beyond the words expressed by the text, it is the overall philosophy behind the text that is important as it expresses a realization and awareness. The philosophy in question is looking for a balance between the two opposing groups of action, developers, in particular for the tourism industry, the conservationists who want the coasts to be preserved and accessible to all in the name of equality for which the State is ultimately responsible. The realization is related to the growing awareness of the precariousness and the fragility of the coastline. Therefore, we should preserve what should absolutely be preserved, depending on the importance of the sites, landscapes, degradation level of the ecosystems and on the limits that should not be crossed. So, marine marshes and sand dune structures will be better, if not absolutely, protected, to name only these examples.

5.4.4.3.2. The French Protection Agency for coastlines and lakeshores (Conservatoire du littoral)

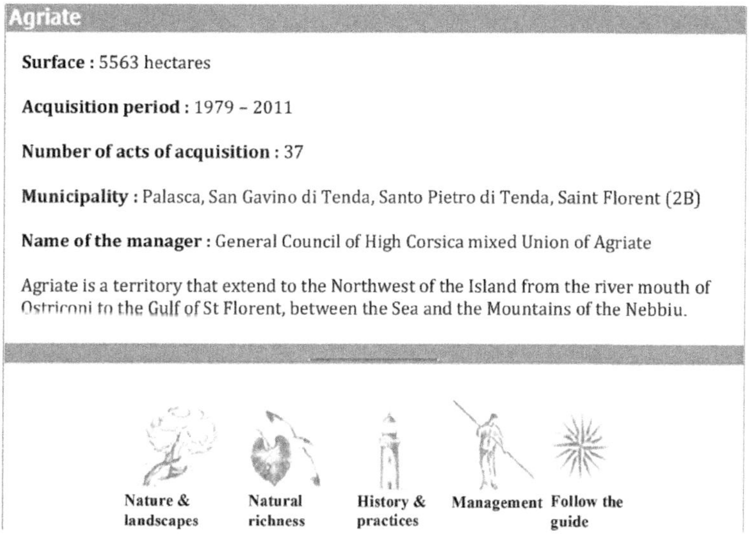

Figure 5.9. *Characteristics of the Protection Agency: the estate of Agriate in Corsica*

The French Protection Agency for coastlines and lakeshores was created by law on July 10, 1975. It was formed as a result of some proposals from the Piquard Report. In the true spirit of a workgroup and the State, it was designed to support what was then referred to the "natural third-party", as the ambition was to preserve at least one-third of the coastal strip and to leave it in a natural state. The report was also produced as the consequences of so many excesses by the anarchical urbanization of the coasts and shores that were being felt. The report was supposed to implement a policy for the "natural third-party or the wild third-party" so as to "conserve part of the coastal area" and therefore advocated for people to buy land without selling it further and to "ensure its management respectful of the practical conditions close to its users". These strong principles mark the origin of the creation of the Protection Agency for coastlines and lakeshores through an act on July 10, 1975. The Protection Agency was (and remains) the strong arm of the State in terms of protection perceived not only as defending sensitive areas, and restoring degraded environments, but also as opening up to the public under the responsibility in particular of territorial collectives. In this, this public body differs from British practices with the National Trust that served as its original inspiration. It is mainly dedicated to the protection of nature and has progressively acquired such a power of influence (indeed, it can be assumed today that most of the territorial collectives adhere to its principles and even anticipate potential requests) that it is an essential tool for conservation, in the more positive sense of the term; heritage conservation that encompasses both nature and all the elements that make up a coastal "landscape" where the presence of man contributes to cultural enrichment. The Coastal Protection Agency is leading the way, and many States are trying to follow the example by asking its experts for advice: from this point of view, the definition of a policy based on heritage values is supported by the existence of a dynamic and respected scientific advisor group. The results of the Protection Agency, at the dawn of the 21st Century, are very encouraging, even though we are still far from the concept of protected "natural third-party" (on January 1, 2011, 138,000 hectares, 1,200 km of coastline and more than 600 sites show undeniable progress). So, the Protection Agency is the tool in France for an ambitious policy traditionally rooted in the action of the State and regulation; it is the result of the implementation of the Coastal law through a few articles and has reinforced the protection (or even the preservation) of some ecosystems and environments both "terrestrial and marine, of remarkable sites and landscapes or characteristics of the natural and cultural coastal heritage" (for instance, Article L 146-6). Figure 5.9 shows an example of the characteristics of the Corsican site of Agriate. The law about the reinforcement of the environmental protection of 1995 has endorsed it in estuaries, so that the debate around the future of large maritime ports is backed by its now recognized authority.

It is a great success story; however, the initial challenge posed is not a foregone conclusion. It could even be argued that it took a long time for the Protection Agency to obtain the authority it deserved among the local collectives with whom it had to negotiate the acquisition of land that was considered as essential to the conservation of the coastal natural and cultural heritage. Once the land was acquired, the areas were to be restored and handed over to municipalities or organizations. As it is steered by an administrative council featuring State representatives and local authorities, the Protection Agency is also a place for permanent dialogue. It appears to be a small structure, only about 100 people in regional delegations symbolically situated in the Corderie Royale of Rochefort-sur-Mer (probably a hint at the great maritime policies advocated by Colvert and a radiant State), it also features about 50 coastal guards recruited by local collectives. Its information panels are inconspicuous but the logo, the blue thistle that disappeared among the sand dunes at the end of 1960, is still visible at the entrance of the offices under light surveillance. The Protection Agency also involves scientific engagement, a bookshop, a clever pedagogy and both savoir-faire and publicization through various channels. Although the Protection Agency was opposed and although it might have given the impression of standing alone in the defense of coastlines, today it appears as the perfect example of an effective and intelligent stakeholder within.

5.4.4.3.3. Marine-protected areas

Marine-protected areas have formed part for a few decades of the framework for ocean management policies with the overarching aim to secure a sustainable development of its activities. Of course, this claim can be argued against along with the enterprise marine conservation, as well as the conservation of all living species, fauna or flora. Indeed, fauna and flora are interdependent. On the one hand, scientists and nature lovers have noticed a progressive deterioration of the quality of the water, with the various forms of pollution affecting coastal fringes, and with the attacks on ocean biodiversity (whose limits are yet unknown). On the other hand, the issue of the extinction of some species makes the question of "sustainable" fishing challenging, irrespective of the causes of these extinctions. Of course, humans are responsible for overfishing; however, everyone is aware that only time and protected areas will allow the populations to recover in high enough numbers to be exploitable by humans. Furthermore, tourism is responsible for part of the degradations, but it is also important to offer tourists not only a high-quality environment but also zones in which they can have the leisure to admire aquatic fauna and flora. This is the logic behind the creation of reserves, which are often very localized, such as the islands or islets harboring coastal avifauna, both migrant and sedentary. The aim behind marine-protected areas appeals to a different dimension, and in order to be effective, the surface area needs to be large enough for the positive effects to be rapidly felt. Most large countries now have marine-protected areas. In the United States, for example, the *National System of Marine Protected Areas* (MPA) was made up in 2009 of 297 sites (4% in water) under various management models, e.g. the federal State, coastal States or even non-institutional partners. All the areas allow all the possible uses, on the condition that these are responsible, which supposes good integration of the partners, through monitoring of the state of the environment on the basis of well-established indicators. The stakes in the end relate to the heritage, whether we are referring to resources or places: historical sites are considered as marine-protected areas. However, the hierarchy of priorities is established for each sites: 91% strictly relate to the conservation of resources and, the most important one, the *Papahanaumokuakea Marine National Monument* to the northeast of the Hawaiian islands is protected purely for conservation purposes. As to France, it aims to make 10% of areas under French law become marine-protected areas and 20% by 2020, with half of these either under fishing restriction. This is an ambitious target that justifies the work of the marine-protected area agency created in 2006.

Marine-protected areas are part of the general policy of ocean protection and protection especially of nature in the oceans. These are the institutional tools with spatial dimensions of sustainable development, as they involve preserving remarkable habitats, protecting biodiversity, restoring fish stocks as much as possible as well as managing the reserves with the possibility for tourists to discover them. The political hierarchy is clear: global level, regional level (for instance, Europe and its various regulations), national level and finally local implementation. So, many projects supported by DATAR (in 2005, known as DIACT) for the launching of the integrated management plan of coastal zones mentioned conservation while allowing people to circulate freely and exercise their profession in the areas in question. Marine nature reserve follows the same logic.

5.4.4.3.4. From land to sea, an initiative of the European Union

Directive 2008/56/CE of the European Parliament and Council, dated June 17, 2008, follows the same line of increasing concern over the ecological health of the seas and adjacent coastlines. It is the latest great directive of the Union and addresses the degradation of marine ecosystems that are the basis of sustainable development. The directive is precise and well documented and is to be applied through to the level of each of the member States for which the directive is relevant, which means almost all of them as they all contribute to telluric and various other forms of pollution being discharged in the watersheds and ultimately released in the sea. The ambition is even greater given, as deadlines are short. The latest is in 2020, to reach a "healthy ecological state". Through its framework directive for the marine environment, the EU takes a stand in the global architecture driven by large international organizations and particularly the environment program of the United Nations. It is now up to France to implement its own strategy in the line of the objectives set by the Grenelles of the Environment and Sea and with the successive stages since the discourse of the French President in the port of Le Havre, in the French marine policy, which was then put into practice by the General Secretariat of the sea.

In practice, the aim of the EU is to promote the sustainable use of the seas and the conservation of marine ecosystems. In order to achieve this, the impact of human activities should be monitored and protected marine areas should be defined. The implementation of a global network of protected marine areas is mentioned, which evokes a network of protected marine areas. The directive refers to the measures taken in the past with good reason, namely with the Habitats Directive and the Birds Directive. Europe is methodically building measures to protect the environment. These measures question some practices (even some State policies), first and foremost with fishing, as it is mentioned several times. The keywords of the directive refer to the state of the environment: namely, marine ecosystems and healthy ecological state. In order to define the health state, some indicators are cited in the appendices, and the member States are required to quickly report on the quality of their waters. Seabed (the issue of fishing methods by trawling transpires here). Furthermore, the first article of the directive urges for the States to "take all the necessary measures to achieve or maintain the marine environment in a healthy ecological state by 2020 at the latest". The areas are marked broadly as the north-east Atlantic, the North Sea (in the broad meaning of the name including the Channel and Kattegat), the Celtic seas, the Gulf of Gascogne (Biscaye) and the Iberian coasts. The "bio-geographic zone of Macaronesia" is actually the only time it evokes a biogeographical region, as if the other ones were not. The States are required to provide a report and to present conservation measures: scientific research is thereby the basis of the policy. It is relatively logical but faces the specialized scientific world with its responsibilities: coherence in practices is not the stronghold of the scientific community, which is often more closed than we would think to the diffusion and discussion of its contributions. However, the directive is implicitly implementing a more integrated management system in which the scientific community can find its place. Life sciences and natural sciences lie at the heart of the writing of the text as appendices I and III list the (11) qualitative indicators to a healthy ecological state along with a very detailed although "indicative" of the forms of pressure and impacts. A healthy ecological state is based upon the marine trophic chain that needs to be better known: it is even the biggest challenge of this directive that innovates in taking broad ecosystems as the basis for the management of the seas. This notion remains in need to be defined more specifically. By targeting human-driven eutrophication (indicator 5), the integrity of the seabed (indicator 6), pollutants (indicators 8 and 9), and finally fish stocks, it is the fishing activity that is in the limelight. This is likely to create challenges for almost all the States, in particular for France where the principles of the framework-directive was added to the code for the environment. At the national level, although the steering is in the hands of the Ministry for Ecology and Sustainable Development (and for the sea), it is IFREMER and the Agency of Protected Marine Areas which are in charge of coordinating the work to be carried out. This is likely to spark debates about boundaries between organisms. The measurement program is under time constraints: the initial evaluation of the ecological state should have been ready by July 15, 2012, the monitoring program is expected on July 15, 2014 so that in 2018, a first assessment of the measures taken can be produced. These ambitions will be judged as excessive.

5.5. Conclusion

The coastline is a changing heritage. The coastline is both land and coastal sea: a situation of interface, reciprocal enrichment, reciprocal fertilization, which unsurprisingly attracts humans and does so increasingly in the context of globalization. Globalization travels on the oceans, uses them as maritime routes and exploits them for its fish stocks but also increasingly for its energy. The "heritagization" is first and foremost the acknowledgment of a principle, i.e. that development can only be sustainable as long as the resources of our planet are managed in a sensible way. In order to achieve this, it was necessary to slow down the growing occupation of the coastline, by distinguishing the most fragile and important environments for future uses, by changing the scale of the analysis. Marine marshes, mangroves, wetlands and coral reefs are all essential for the protection of future fish resources. It is in the coastal areas that these protected areas can be found, easier to define on land than at sea. These areas are in the custody of the French "Coastal Protection Agency" (*Conservatoire du Littoral en France*) on the one side, and "Marine Protected Areas" on the other side, at the global scale. The protection of nature is a political acknowledgment in the full meaning of the word of this heritagization: it is apparent that the effort has been constant for at least three decades and that beyond the display of principles, it is of course in the field that the actual protection is moving forward. This in spite of the pressure, on the one hand, of those who think it is going too fast, and on the other hand, of those for whom it will never go fast or far enough for that matter. Overall, the principle of "integrated" coastal management offers the framework for a debate between user groups with antagonistic interests that are supposed to converge. This is the price for sustainable development.

5.6. Bibliography

[BIR 85] BIRD E.C.F., *Coastline Changes, A Global Review*, John Wiley, Chichester, 1985.

[BIR 93] BIRD E.C.F., *Submerging Coasts, the Effects of a Rising Sea Level on Coastal Environments*, John Wiley, Chichester, 1993.

[BUH 09] BUHOT C., *Tensions foncières sur le littoral*, Presses Universitaires de Rennes, Rennes, 2009.

[CAR 10] CARIOLET J.M., COSTA S., CASPAR R., *et al.*, "Aspects météo-marins de la tempête du 10 mars 2008 en Atlantique et en Manche", *Norois*, no. 215, pp. 11–31, 2010.

[CIC 98] CICIN-SAIN B., KNECHT R.W., *Integrated Coastal and Ocean Zone Management, Concepts and Practices*, Island Press, Washington, 1998.

[COC 91] COCH N.K., WOLFF M.P., "Effects of hurricane Hugo storm urge in coastal South Carolina", *Journal of Coastal Research*, Special Issue no. 8, pp. 201–226, 1991.

[COU 12] COULIOU J.R., "Entrées stocks, surpêche", in MIOSSEC A. (ed.), *Dictionnaire de la mer et des côtes*, Presses Universitaires de Rennes, Rennes, pp. 472–273, 2012.

[DES 09] DESPRES L. (ed.), *L'estuaire de la Loire, un territoire en développement durable?*, Presses Universitaires de Rennes, Rennes, p. 470, 2009.

[EMS 10] EMSAGG, (*European Marine Sand and Gravel Resources*), *Journal of Coastal Research*, special issue 51, West Palm Beach, 2010.

[FEU 12] FEUILLET T., CHAUVEAU E., POURINET L., "Xynthia est-elle exceptionnelle? Réflexions sur l'évolution et les temps de retour des tempêtes, des marées de tempête et des risques de surcotes associés sur la façade atlantique française", *Norois*, no. 222, pp. 27–44, 2012.

[GOA 10] GOASGUEN G., "Aspects météo-marins de la tempête du 10 mars 2008 en Atlantique et en Manche", *Norois*, no. 215, pp. 11–31, 2010.

[GOU 10] GOUYON P.H., LERICHE H., *Aux origines de l'environnement*, Fayard, Paris, 2010.

[GOU 12] GOULLETQUER P., *Biodiversité et environnement marin*, Editions Quae, Versailles, 2012.

[GRA 07] GRATALOUP C., *Géohistoire de la mondialisation, le temps long du Monde*, A. Colin, Paris, 2007.

[GUN 09] GUNNEL Y., *Ecologie et Société*, A. Colin, Paris, 2009.

[LAV 11] LAVIGNE F., PARIS R., *Tsunarisque, le tsunami du 26 décembre 2004 à Aceh, Indonésie*, Publications de la Sorbonne, Paris, 2011.

[LEA 79] LEATHERMAN S.P., *Barrier Islands: From Gulf of St. Laurence to Gulf of Mexico*, Academic Press, New York, 1979.

[LEU 13] LEULIETTE E., The Budget of Recent Global Sea Level Rise, 2005-2012, 2013.

[MER 09] MERCKELBAGH A., *Et si le littoral allait jusqu'à la mer! (la politique du littoral sous la Ve République)*, Editions Quae, Versailles, 2009.

[MIO 93] MIOSSEC A., La gestion de la nature littorale en France atlantique. Etude comparative (Royaume-Uni, Pays-Bas, Espagne, Etats-Unis), Doctoral Thesis, University of Brest, 1993.

[MIO 98a] MIOSSEC A., "L'estuaire de la Loire, un archétype des problématiques estuariennes entre développement économique et protection de l'environnement", in GAMBLIN A. (ed.), *Les littoraux, espaces de vie*, SEDES, 1998.

[MIO 98b] MIOSSEC A., "Rythmes et bilans dans les domaines littoraux", in *L'érosion entre Nature et Société*, Chapter 6, SEDES, 1998.

[MIO 99] MIOSSEC A., "L'aménagement du littoral français entre développement économique et protection de l'environnement", in *Le Littoral, principes et méthodes*, Editons BRGM, Orléans, 1999.

[MIO 09] MIOSSEC A., *Les littoraux entre Nature et Aménagement*, A. Colin, Paris, 2009.

[MIO 12] MIOSSEC A. (ed.), *Dictionnaire de la mer et des côtes*, Presses Universitaires de Rennes, Rennes, 2012.

[PAS 81] PASKOFF R., *L'érosion des côtes*, Que Sais-Je? PUF, Paris, 1981.

[PAS 93] PASKOFF R., *Côtes en danger*, Masson, Paris, 1993.

[PAS 98] PASKOFF R., *Les littoraux, impacts des aménagements sur leur évolution*, Masson, Paris, 1998.

[PIN 74] PINOT J.P., *Le Précontinent Breton entre Penmarc'h, Belle-Ile et l'escarpement continental, étude géomorphologique*, Impram, Lannion, 1974.

[PIQ 74] PIQUARD M., Littoral français, perspectives pour l'aménagement, Internal Report, 1974.

[PIR 96] PIRAZZOLI P.A., *Sea Level Changes: the Last 20000 Years*, John Wiley, Chichester, 1996.

[TOU 12] TOUSSAINT J.F., *L'homme peut-il s'adapter à lui-même. Marges d'adaptation de l'espèce humaine face aux changements environnementaux*, Editions Quae, Versailles, 2012.

[VAN 77] VANNEY J.R., *Géomorphologie de la marge continentale sud-armoricaine*, SEDES, Paris, 1977.

[VIN 12] VINET F., DEFOSSEZ S., REY T., *et al.*, "Le processus de production du risque "submersion marine" en zone littorale, l'exemple des territoires "Xynthia"", *Norois*, no. 222, pp. 11–26, 2012.

6

From Vulnerability to Adaptation to Climate Change: Food for Thoughts in Social Sciences

6.1. Introduction

This chapter does not aim to give the exhaustive state of the art of scientific studies around issues of vulnerability and adaptation to climate change. The objective is instead to provide a global approach to both concepts which would enable social science researchers, both experienced researchers and beginners and those coming from various disciplines, to develop fieldwork methods and boost knowledge.

In order to clarify the meaning of the concepts of vulnerability and adaptation and to specify their main challenges, this chapter will be broken down into four sections. Figure 6.1 summarizes these with the theme being the progress of vulnerability and adaptation mechanisms.

Section 6.2 not only briefly presents the origin of the concept of vulnerability, but it also explains how the vulnerability of a given society lies within the various interconnections between, on the one hand nature, timeframes and the scale of the natural hazards, and, on the other hand, the temporal and spatial dimensions of the place.

Chapter written by Alexandre MAGNAN.

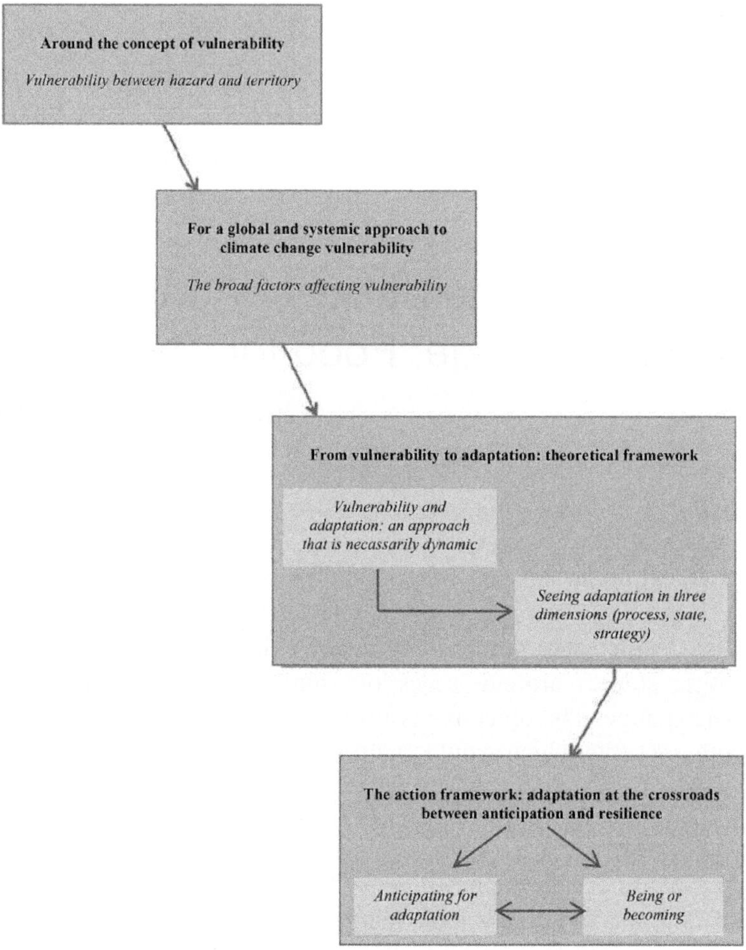

Figure 6.1. *General structure of the chapter*

Building on these bases, section 6.3 develops the hypothesis according to which the present level of vulnerability is rooted in historical characteristics proper to the shaping of the society and its territory. This section provides a global and systemic approach to vulnerability through six broad factors of influence: the spatial configuration, the sensitivity of the ecosystems, social cohesion, the diversification of the economic and subsistence activities, the political and institutional organization, and the living conditions.

Section 6.4 then links the vulnerability and adaptation concepts to climate change. It first of all shows that both these concepts supply one another: indeed being able to adapt helps in reducing vulnerability, which in turn favors the ability to adapt. These are therefore dynamic concepts. The section then goes into the detail of the concept of adaptation and identifies three main dimensions (process, state and strategy).

Finally, the last section (section 6.5) departs from the theoretical framework to enter a more practical one. It discusses the principle of flexibility as the fundamental characteristic of a society which enables it to go beyond the barriers imposed by uncertainty around the local impacts of climate change. Two key-pillars are put forward: anticipation and resilience.

6.2. Around the concept of vulnerability

The relationships between the concepts of risk, vulnerability, hazard and territory are complex enough to deserve our attention. Although hazards related to climate change will hold a priviledged position here, other phenomena unrelated to marine-meteorological conditions (e.g. earthquakes) will be cited occasionally so as to explore the different dimensions of risk and vulnerability.

The first researchers who became interested in risk were natural scientists, as could be expected due to the nature of the hazards themselves (volcanic eruptions, earthquakes, cyclones, etc.). Then progressively, other scientists opened up to the human aspects of disasters. The cumulation of analyses introduced the idea that there is a difference in terminology between risk and disaster, as the first term refers to the probability for the occurrence of a hazard, whereas the second refers to its actual consequences on the system [BLA 94, DAU 01, VEY 07]. This led to the widening of horizons around risk factors [BEC 01, CAR 04], which sets the necessary foundations to the notion of vulnerability. The latter was first defined as the simple level of exposure to risk, then it progressively increased in reputation to finally become a scientific concept in itself [ADG 06, BAN 04].

6.2.1. *The development of vulnerability concepts*

On these grounds, the two main visions of risk, and indirectly of vulnerability became more prominent between 1950 and 1980, which favored the emergence of the third concept. Today, the third concept is the one which is most clearly defined of all. There was also a gradual realization of the significance of interactions between humans and the space they inhabit [FRA 03, HIL 04, REG 06]. These interactions are not limited to the exploitation of natural resources. On the contrary, they go beyond this relation through the relationships related to history and identity with places and, by extension to the role of the environmental characteristics on the dynamics and attitudes of the various societies [DIA 97, DIA 05]. Going beyond the man/environment dichotomy enabled researchers to introduce more complexity in the approach to natural phenomena and their consequences on societies [BAN 04, OBR 04]. From an epistemological point of view, the change consisted of overturning the domination relation between nature and society when explaining a disaster, which can be shown by going back to the three paradigms (physical, structural and complexity[1]) developed by D. Hilhorst [HIL 04].

The *physical paradigm* approach emerged in the 1950s. It relied on the idea that the causes of a disaster were to be found in the physical characteristics of the hazards and of the impact process. Hard sciences (geology, climatology, etc.) were then dominating the field of study around natural hazards; whereas the nature of societies only played a marginal role as an explaining factor. The role of social sciences was then confined to the identification of individual behaviors when faced with risk and disaster (*a posteriori* analysis). This approach neglects the intrinsical characteristics of societies when explaining a disaster. This idea became popular during the 1980s through the work of anthropologists, geographers and sociologists. The approach changed from 'hazard-focussed' to a structured school of thought around the *structural paradigm* with at its center, the idea that 'Disasters were not primarily the outcome of geographical processes.' ([HIL 04, p. 53],

1 D. Hilhorst refers to the behavioral paradigm, the structural paradigm and the complexity paradigm.

quoting K. Hewitt[2]). The role of poverty was mentioned among other factors and through it, the social and economic processes associated with it. So the socio-cultural dimension of vulnerability to natural hazards emerged and imposed the need for a differentiation between the physical processes (hazards) and the human processes (vulnerability). The traditional formula 'Risk = Hazard * Vulnerability' incidentally arose from the development of this concept. The neat separation between physical and human-related processes however turned out to be unsatisfactory in that it does not lead to the understanding of why the various groups within a single population experienced the impacts of the same hazard differently; in other words, why did they exhibit different degrees of vulnerability.

Therefore, a third concept appeared in the 1990s, which led to the *complex paradigm* or the *mutuality paradigm*. In this case, the mutual aspect of physical and human processes was emphasized thereby pointing out that if the hazard exerts any direct influence on the way a society functions, human activities in turn affect the probability of a hazard being triggered, in other words of a disaster occurring. At a local scale, this principle of mutuality is particularly evident when looking at issues of coastal erosion. Although fighting this phenomenon, which is first and foremost inherent to the natural shortage in sediments, requires the deployment of groynes and seawalls, perversely, the observed effect of these defense strategies is the worsening ability to maintain sand in the medium-term, which reinforces the initial erosion problem. On a global scale, human activities reinforce the greenhouse gas concentration in the atmosphere, which increases temperatures in the lower atmosphere and the surface of marine water on the one hand, and an acceleration of glacier melting on the other hand. These phenomena act in combination to explain sea level rise. Both examples show that the physical processes (the hazards) and the human processes affect each other more than if they only interacted at any one particular moment in time, i.e. when the hazard occurred [BLA 94]. Although the mutuality paradigm undeniably made a breakthrough in enabling the concepts of risk and vulnerability to move forward, it also made

2 [HEW 83].

the process of identifying pragmatic strategies to reduce risks a lot more complex [WIS 04]. This is true firstly as the number of variables to consider has been increased (with an addition of variables of different natures), then as it requires the consideration of both the interactions among these variables and their respective latencies.

Indeed, although the impacts of a hazard are first and foremost direct impacts (human losses, various damage, breaking of networks, etc.), they can also expand over time following the domino principle [DAU 07, PRO 07], thereby extending the 'lifespan' of a disaster. This is all the more true in the context of climate change, which compels us to take time into account (gradual risks and related uncertainties). By reducing the timescale, it can be noted that the vulnerability of a given space plays an important role in the way the consequences of a disaster unfold. Indeed, the nature of this type of vulnerability is twofold: it first touches on the fragility (environmental and/or human) of a system faced with a hazard, and then on the capacity of this system to absorb the crisis and find a balance after it is over (concept of *resilience*). Fragility can be used to explain the most direct impacts, whereas resilience is used to understand the impacts that occur through the domino effect. So how can resilience be analyzed without integrating the various interactions among the components of the society in question on one side, and between the society and its environment on the other? The relevance of this question is based on the mutuality paradigm which forces us to adopt a systemic and dynamic vision of the human/environment relationship and therefore of risks, vulnerability and hazards. In terms of the analysis, vulnerability needs to be considered as the result of both endogenous and exogenous influences [DUV 14, FUS 06], which the structural paradigm failed to specify; whereas the level of vulnerability needs to be understood as evolving in time.

A relatively well-thought through definition of vulnerability emerged at the beginning of the 21st Century as a consensus was reached among researchers from around the world united within the IPCC[3] on the issue of climate change. The report explicitly refers to the human/nature interface and to relationships among humans. In the

3 Intergovernmental Panel on Climate Change (IPCC) – www.ipcc.ch.

IPCC report, the notion of vulnerability is defined as "the level at which a given system is at risk of experiencing or being negatively affected by the negative effects of climate change, including climate variablity and extreme events. Vulnerability depends on the nature, the scale and the rhythm of climate changes to which the system is being exposed as well as on its sensitivity and its adaptive capacity". Although this definition is not exempt from ambiguities [BRO 03, HIN 11, MAG 12], which again emphasizes the complexity of the topic, it has the advantage of indirectly characterizing vulnerability through a triple dimension: nature, scale and rhythm.

6.2.2. *Hazard, place and vulnerability*

At this stage, it is crucial to understand how notions of perturbation[4] (hazards) and of place come together to express a certain level of vulnerability. Perturbation and place can hence be described following the three dimensions expressed by the IPCC (nature, scale and rhythm) and show the different overlaps that define vulnerability (see Figure 6.3). Table 6.1 shows the equivalence of the terms used henceforth.

IPCC	Traditional definition of hazard/event/perturbation	Equivalence terms within this study	
		Hazard	Territory
Nature	*Nature*	Nature	Nature
Rhythm	*Frequency*	Time frames	Time frames
Scale	*Intensity, place*	Scale	Spatial scale

Table 6.1. *Equivalence of the terms used in the demonstration below (to be looked at in conjunction with Figure 6.3)*

4 In this case, the term perturbation is chosen over hazard as it is easier to understand in most social sciences as, apart from geography, they do not involve any specific training in the physical mechanisms responsible for the hazard in question. So this generic term seems more appropriate to enable social sciences to integrate both one-off and gradual environmental changes, and therefore to take into account combinations of hazards and risks.

6.2.2.1. *Hazard*

The *nature* of a hazard is often very complex. If it originates in a particular environment, its expression as a physical process (winds, swells, landslides, etc.) can occur within one or several of the three great spheres air/land/sea. This is the case for cyclones, which start over the ocean and generate on the coasts both violent winds and strong swells with heavy rain. Volcanoes can also combine lava flows, pyroclastic flows and landslides. So beyond the original nature of the hazard (the cyclone, the volcanic eruption), it is mostly the unfolding of physical processes which needs to be considered to account for the transformation of an original hazard into a series of multiple secondary hazards. The need for the combinatory approach is even more pressing when considering the effects of climate change. Indeed, the rise in temperatures is expressed over the long term by the acceleration of glacier melting and an increase in the volume of oceanic mass (thermal dilation process), i.e. by a rise in sea levels which, when associated with other processes, leads to the exacerbation of the the impacts of extreme events (e.g. storms) and of more gradual climatic changes (e.g. salinization of groundwater and of soil). The challenge with climate change is therefore to be able to take into account all the various changes in the natural processes. This explains why we refer to the term hazard here in its very broad meaning, which includes both the processes and their one-off events (hazards).

From the point of view of timeframes, hazards can be either isolated in time (e.g. violent thunderstorms, floods or even forest fires) or diffuse and gradual. The issue of climate change combines both mechanisms as the possibility of intensifying and exacerbating more or less recurrent one-off events (droughts, heat waves, cyclones, etc.) is an indication for more gradual changes both in terms of conditions (temperatures, currents, precipitation, etc.) and in terms of resources (water, fish, etc.). Climate change is therefore associated with a more profound change of natural environments and indirectly, of the contexts in which human societies develop. This introduces another important element: the timescales of some hazards may be complex, half-way between discrete in time and gradual. This is typically the case for drought events whose incidence can vary from a few days to several months, or even years, and which can result in a latent water shortage that would be revealed by a lack of precipitation, by higher temperatures than normal and/or by overconsumption problems.

Timeframes affect the nature and scale of the impacts, and therefore of the forms vulnerability can take. On the basis of this assumption, N. Brooks [BRO 03] proposed a classification of hazards following three categories. Namely the 'discrete recurrent hazards', first, which includes short hazards repeated in time, with lower or higher frequencies. Storms, drought events or floods typically feature in this category. These types of hazards are relatively well understood today, as scientific research has looked at case studies and at the analysis of their various impacts. The second category, that of 'continuous hazards', refers to continuous phenomena such as the increase in mean temperatures on the Earth's surface, sea level rise or desertification. Modern societies are unarmed when faced with these hazards, as although technical solutions may exist to prevent discrete hazards[5], addressing continuous risks requires profound behavioral changes and changes in development strategies, which assume long-term projections [MAG 13]. This is incidentally what explains the main fears when it comes to climate change. Finally, the third category regards 'discrete singular hazards', i.e. phenomena which only occur on a one-off basis as opposed to recurrently, and in a brutal or progressive way. Their identification depends on the timescale considered: climate-related stress[6] or a change in ocean current circulation for instance. Of course, this marks the limit between the two previous categories but it enables reinforcing the idea that risk and therefore the vulnerability level are understood differently according to the time step considered.

Finally, hazards are defined by their *scale*, in this case, i.e. by their spatial footprint. The tsunami in Sumatra (December 2004) for instance was felt all the way to the Eastern coasts of Africa which were several thousands of kilometers away from the source of the

5 Solutions to these hazards are often costly and associated with variable efficiency which do not depend only on technical aspects (acceptance by societies for example).

6 Here, we enter the sphere of distinction between climate variability and climate change.

tsunami. Other phenomena can be extremely localized such as fires or marine submersion events. The expected sea level rise or global warming are global problems due to the nature of their impacts, although they do not have the same effects everywhere. The relevance of the spatial dimension(s) of the hazard is therefore two-fold. First of all, it resides in the understanding and identification of the processes and potential impacts, including through mapping of the exposed zones; then, it lies in the identification of adequate measures to be taken in order to limit risk. One of the most topical questions the tsunami raised in December 2004 regarded the implementation of an alert system that would be common to all the countries surrounding the Indian Ocean. In a country such as the Maldives, where 90% of its 298 km^2 land surface is situated at less than one meter above the sea level, local anticipation measures are limited[7]. However, flood management for instance does not often require technical solutions of international scale and their regulation can be achieved at a local (system of prevention and aid) to a national scale (insurance systems for instance). The challenge of climate change consists of combining international, national and local scales, through the mitigation of greenhouse gas emissions and through adaptation strategies with a common goal, namely limiting the rates of disasters.

6.2.2.2. *Territory*

Territory also responds to the three dimensions of nature, timeframes and spatial scales. Its *nature* confers specific characteristics which relate both to its spatial configuration, expressed in terms of potential/constraints, and to the society it supports and which defines exploitation strategies for its original potential/constraints. A large continental surface made up of large fields and covered homogeneously in numerous water bodies does not offer the same potential for agricultural development than an oceanic archipelago or a dry Mediterranean island. Similarly, two environments with comparable spatial configurations can in some cases exhibit neither the same axes of development (e.g. tourism

---

7 Which is not stopping the country to try and implement innovative solutions today [DUV 12].

rather than agriculture), nor the same level of development due to differences in terms of historical heritage, political choices or levels of exchanges with other places [DIA 97].

In parallel, a territory, even if it is defined first and foremost as the product of systemic links between a community and its environment [COL 99, DIM 91], is made up of a number of territories with inferior hierarchies, of a mosaic of environments which function as micro-territories. As a result, it is extremely difficult to define territory spatially, whatever its size, as a system is always simultaneously composed of subsystems and part of a wider system [PEG 96]. So there are several *spatial scales* to territory (see Figure 6.2).

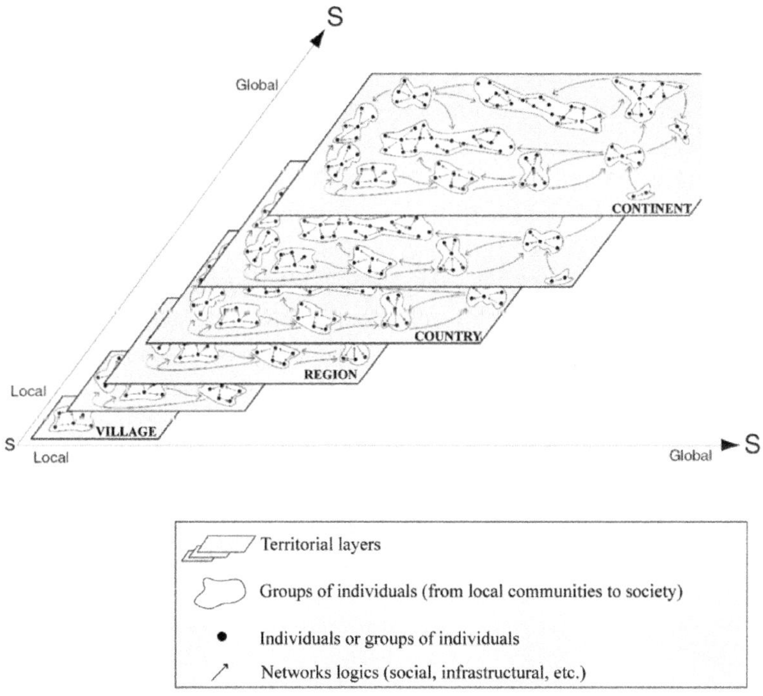

Figure 6.2. *Spatial components of territory*

Finally, the characteristics proper to the place in relation to the groups inhabiting it, as well as the interactions between the spatially-defined territory and its neighbors, define the ways in which the territory is built and evolves. We are here in a logic of process, and therefore in a dimension of *timeframes*. The latter is crucial for the good understanding of the territory, whose current potential and constraints result from socio-cultural, economic, political and environmental changes more or less recent in time and more or less rapid. The present is rooted in history; the vulnerability of a place faced at any given point in time, with a hazard, is characterized by all this temporal depth; furthermore, it varies from one scale to another and it depends on characteristics proper both to the hazard itself and to the territory.

6.2.2.3. *Vulnerability, where hazards meet territory*

When comparing the three dimensions of hazard on the one hand, and of territory on the other, a conclusion emerges; namely that the relation between the strengths and the limits of a territory subject to a natural hazard, which explains the scale and nature of the impacts, is intrinsically linked to the relations among the six variables (Figure 6.3). The scale of the hazard for instance influences the temporal aspects of the territory as it is expressed through a surface impact of larger or smaller dimensions which encompasses in fact a larger or smaller number of zones with economic activities. This in turn affects the rhythms of the development process. As an example, let us note that although the intensity of the tsunami wave was stronger in India and Sri Lanka than in the Maldives, the economic repercussions of the disaster in 2004 were proportionally less significant in the first two countries than in the Maldivian archipelago [COS 05, LAN 05]. There are two reasons behind this: India and Sri Lanka have a shorter coastal strip[8], so *a priori* and when respecting all the proportions, there is a smaller exposed surface area but more importantly, on the one hand the activities at the heart of the economies of these two countries were not situated in the devastated coasts and on the other hand, these countries were in relatively high

---

8 Ratio between the length of the coastal strip (in km) and the total surface area of the country (in km^2).

growth phase. In the Maldives on the other hand, tourism is exclusively related to beach resorts and contributes to a third of the GNP and to three quarters of foreign currency influx. If we look at the situation at a finer scale, all the Maldivian islands did not react in the same way to thc tsunami wave both in terms of damage and in terms of population evacuation. Differences were very striking between islands which were only distant less than one kilometer [MAG 06, DUV 14]. So the vulnerability level varies in space sometimes at a very fine scale.

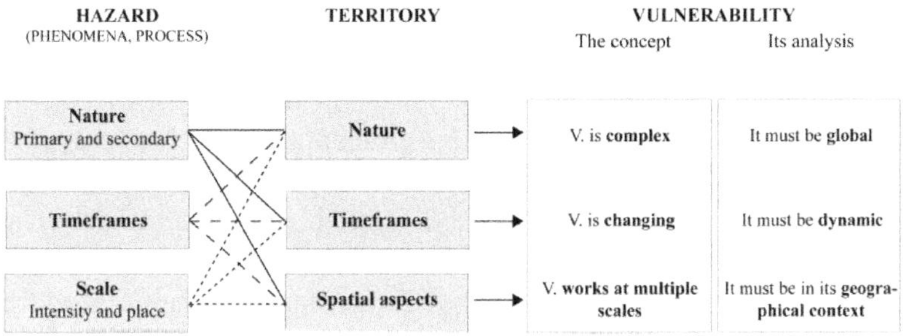

Figure 6.3. *Spatial components of the territory*

On a different level, the timeframes of the hazard have an effect on the timeframes of the territory, more precisely on spatial planning. Indeed, hazards are expressed as a threat which are either limited in time, or gradual, so it is more or less pressing in terms of reducing risk and of planning space. This can bring a given society to settle on the coasts protected from recurrent storms but situated in a region which will be subject, in a few decades, to problems of groundwater depletion, as the latter will be poorly-replenished due to a flawed hydrological regime. In this case, risk in the short term is usually seen as more important in terms of public policy than the risk in the longer term.

The case of the creation of new tourist attraction poles on mediterranean coasts is a perfect example of the limited ability of planners and economic operators to take the long term into account.

This is the case even though it is known that climate change, in the current state of knowledge and when taking into account the weight of local factors in regional developments (at the scale of the Mediterranean), raises *a priori* both constraints and opportunities [PER 01]. In Tunisia, for instance, a tourist project (fifteen hotels, a golf course, and a marina) involves the partial filling of a laguna (*Lella Hadhria*) on the Eastern side of the island of Djerba, while the sand beach that separates it from the sea is of modest dimensions (a few meters in length, and a maximum of 250 m in width) and in spite of coastal erosion being very active on this part of the coast [OUE 04, PAS 85]. However, following the evolution of coastal currents, sea level rise and its effects, this erosion phenomenon is at risk of intensifying in the coming decades. Why then was this project not designed on (shallow) slopes directly to the west of the laguna? This would have avoided the risk of erosion and the costs associated with it, but it would also have provided future customers with a remarkable landscape (low beach dotted with palm trees and surrounded by a rosy to blueish water). If one of the reasons could have been a very high probability of depleting hydrological resources in this sloped zone – with, for example problems incurred from soil instability – surveys in the field[9] showed that this scenario was not even considered and that on a broader note, developers have a selective vision of the challenges to be taken into account in the medium- and long term. So potential problems related to sea level rise or to water depletion, heatwaves or a change in tourism patterns in Europe are not being taken into account. From a general point of view, the issue of climate change is failing to be taken into account and planning decisions are being made around problems judged as more pressing (economic development for instance). Gradual risks thus seem to be weaker drivers in terms of public and private decisions, either for individuals or collectives, as one-off risks.

So at this stage, we gather that what explains vulnerability is the link between the nature, timeframes and scale/spatial scales of

9 Carried out in July 2009.

the hazard and of the territory. In this sense, the adoption of a global definition of vulnerability by the IPCC marks a crucial breakthrough.

6.3. Toward a global and systematic approach to climate change vulnerability

Climate change is certainly not the only changing factor affecting future living conditions. The March 2011 tsunami in Japan stresses the fact that marine-meteorological risks are not the only ones that should be considered, on the contrary, natural risks in general and their potential combinations should be monitored. On the other hand, climate change is likely to mainly exacerbate pressures which already exist today, such as challenges in achieving high-quality water or the degradation of forest ecosystems for instance. This realization voices the need to open up the approach to risk which today still tends to divide specialists in natural hazards from specialists in the impacts of climate change. These two scientific communities use different terms to explain similar questions. The model we propose here aims not to position ourselves in any of these two communities, but rather to highlight a common vision of the broad factors which are likely to affect the vulnerability of a place and of a system. These broad factors along with their interactions give unifying directions for research which invite to develop cross-disciplinary research strategies in social sciences.

6.3.1. *From vulnerability to general environmental changes*

6.3.1.1. *What the IPCC says about vulnerability to climate change*

We know from the IPCC's definition of vulnerability to climate change (section 6.2.1) that the latter depends on three general components: exposure, sensitivity and the capacity to adapt. The IPCC also provides definitions to these three components. Exposure characterizes "the nature and degree to which a system is exposed to significant climatic variations". Sensitivity is the "degree to which a system is affected, either adversely or beneficially, by climate-related stimuli. The effect may be direct (e.g. a change in crop yield in response to a change in the mean, range, or variability of temperature) or indirect (e.g., damages caused by an increase in the frequency of coastal flooding due to sea-level rise)". Finally, the adaptive capacity represents "the ability of a system to adjust to climate change (including climate variability and extremes)".

In traditional approaches to risk, vulnerability is a function of the hazard and of the level of exposure of the system to the hazard in question. The notions of sensitivity as capacity to adapt are not explicitly mentioned, although they implicitly constitute elements of the analysis. So specialists in resilience tend to refer to the coping capacity of a system, whereas specialists in adaptation to climate change will use the terms adaptive capacity. However, these two notions refer back to similar concepts of the capacity (or lack thereof) of a system both to respond to a hazard and to anticipate the successive hazards. It therefore appears that a "simplified" approach, in that it does not enter conceptual debates opposing specialists in risk and specialists in climate change, is now necessary in a wide social sciences perspective. "Simplified" however does not necessarily mean 'simplistic' as it does not neglect the need for depth in the analysis required by the study of vulnerability. The depth of analysis is justified by the fact that the vulnerability of a system, current and future, is rooted in a certain length of time.

6.3.1.2. *From the roots of society to current living conditions*

Blaikie, Cannon, Davis and Wisner developed at the beginning of the 1990s the "Pressure and Release" model (the PAR Model, Figure 6.4). The basic idea relies on the realization that a "(…) a disaster is the intersection of two opposing forces: those processes generating vulnerability on one side, and physical exposure to a hazard on the other" [BLA 94, p. 22]. Their study mainly highlights the interest in developing a model of vulnerability as rooted in the intimate bases of the society under study. The human factors responsible for the weaknesses of a group can be categorized in three dimensions, which encourages researchers to base their analysis on the fundamental values of the society.

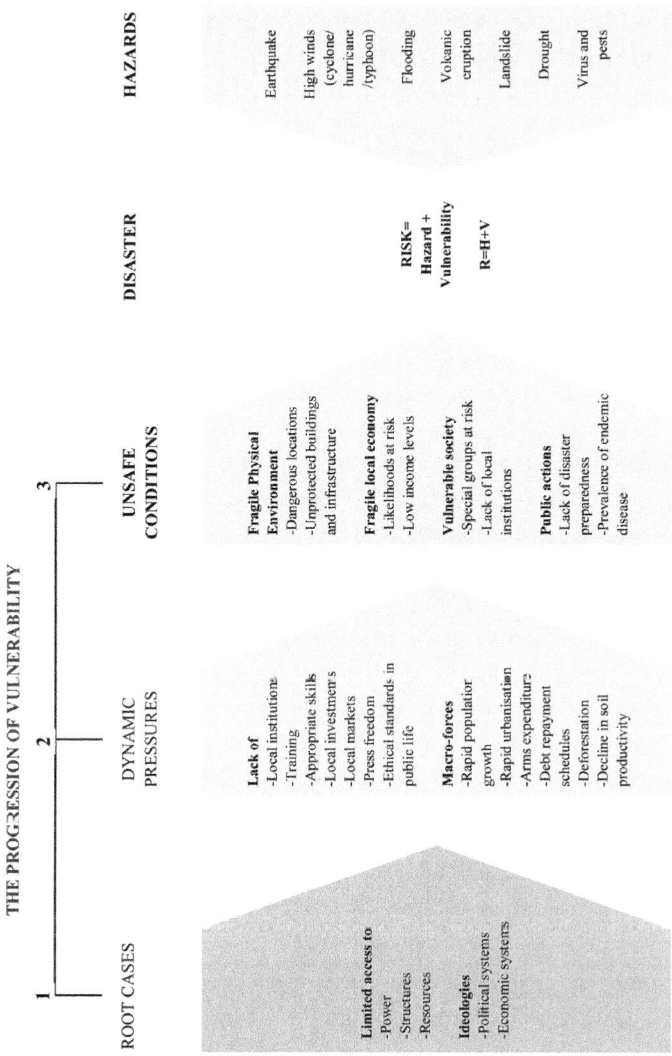

Figure 6.4. *The Pressure and Release model (from [BLA 94])*

The first dimension is related to the characteristics of a human system when the hazard arises. The latter can indeed highlight latent flaws in terms of the electricity networks and/or communication networks. It can also reveal a certain level of fragility of the environment due to the human activities, economic and/or political instability, widening gaps between social groups, etc. Conversely, the conditions prior to the hazard can confer to the system a relative solidity/stability which enables it to be better prepared for the crisis. These *surface characteristics* (as opposed to deeper causes, see below) are situated directly at the interface of the event, the confrontation explains the triggering of a disaster and its scale. If domino effects play a role and if the extent of the disaster in time goes beyond the simple confrontation between the system and the event, it is because other forms of logic underlie the surface characteristics.

These forms of logic relate to the second dimension of the PAR model, i.e. that of *dynamic pressures* in Blaikie, Cannon, Davis and Wisner's words. We are no longer talking about a static state at this stage, but about changes. Taking into account environmental, socio-cultural, politico-institutional and economic dynamics requires that surface characteristics be seen as the result of multiple changes. This implies that the degree of vulnerability which characterizes the system under study is dynamic and was built over time, so any attempt at reducing it (by improving the management of crises or by anticipating them) requires groundwork on the system itself and not just adjusting the surface. As an example, it is the relationship to coastal environments which needs to be thought over again rather than simply fighting against erosion or marine flooding. So these elements appear as parts of a more complex puzzle. Here the occurrence of a disaster is rooted deeper than the simple interface between a system and the hazard and goes beyond the near past (which accounts for the current characteristics of the system) to go back in time until a further past. We go from decades to centuries. The topic of climate change invites us to look at timescales that extend further than what we are comfortable with, as it requires changes in lifestyles and development paradigms. In this sense, it is a real scientific opportunity to test and partially reinvent the development trajectories of our societies.

The third dimension which appears in the PAR Model is rooted even further in time and relates to *root causes*. The timescale, of one to several centuries, refers to the fundamental values on which societies are built: the hierarchy, distribution of power, relation to natural resources, belief systems, economic orientations, etc. These ancestral models through the channel of time and dynamic pressures are at the origin of the recent choices of the group and, by extension, of its vulnerability to hazards.

This multi-dimensional approach to vulnerability leads to the analysis of the factors which affect the level of vulnerability, either by decreasing or by reinforcing it. The objectives are twofold: accounting for the complexity of the vulnerability concept and proposing a theoretical framework. Six generic factors can be differentiated which fall under all three dimensions of the PAR Model as well as those highlighted in the definition by the IPCC (exposure, sensitivity and adaptation capacity).

6.3.2. *The six broad factors affecting vulnerability*

Whether we want to apprehend vulnerability to natural hazards in general or to climate change, the following conclusion rapidly arises: vulnerability is dependent on many parameters (environmental, economic, socio-cultural and/or politico-institutional) and on the combination of these parameters [BAN 01, BEC 12, DUV 14]. Six broad factors affecting the level of vulnerability to climate change and, more broadly, to natural hazards can hence be identified (see Figure 6.5).

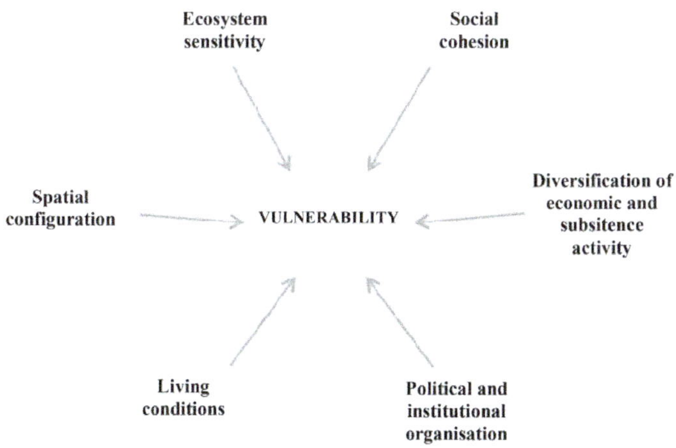

Figure 6.5. *The spatial components of territory*

6.3.2.1. *Spatial configuration*

In practice, spatial configuration manifests itself in terms of potentialities and constraints to the development of societies. It covers two main dimensions.

The first dimension relates to the regional context to which the territory belongs and which will determine a series of potential events, such as cyclones, and the monsoon in tropical regions, or snow storms in colder areas. It sets a general framework in which natural hazards occur.

Secondly, all the areas found in the same climatic region (intertropical, temperate and polar) are not subject to the same hazards as these vary in nature as well as in intensity, and as their impacts are not felt uniformly everywhere. The type of coast, at the local and micro-local scales, for instance, plays a crucial role: the lidos in Languedoc and in Camargue (France) will not respond in the same way to high waves than the rocky shores around Greek islands. Similarly, the proportion of shoreline and the level of ramification of the hydrographical network, the underwater morphology, etc., all act in conjunction to explain the level of exposure of the given territory to hazards (submersion and floods, for instance).

6.3.2.2. *Ecosystem sensitivity*

The ecosystems making up a territory are varyingly disturbed by changes in conditions (sparser rain, stronger winds, warmer waters, etc.). We refer here to their sensitivity which undisputably influences the probability for a hazard to affect or not the balance in place. This will affect, in the long run, the vulnerability of the whole territory. Fauna and flora can indeed play a role in reducing risks, as is the case among others with coral reefs, mangroves or sand dunes which, on the shorelines, are one of the buffer areas against swells and storms. These natural buffers can however be affected by waves that have higher intensity than normal, or in the case of coral reefs, by the warming of surface waters and/or their gradual acidification. By the same token, plant species that are very sensitive to changes in rain patterns will no longer retain the topsoil in cases of prolonged drought events, which leads to the erosion of the soil and to landslide with the first rainfall.

Therefore, ecosystems are likely to lose their protective function when they become stressed by extreme events or gradual changes, whether these pressures are related to climate change or to human activities in general (pollution, mechanical degradation, etc.).

6.3.2.3. *Social cohesion*

Social cohesion relies on social relationships among the individuals of a group, on the sharing of a cultural identity and, *a priori*, on a relative level of homogeneity among the different economic classes[10]. It can contribute to reducing the vulnerability of a territory by favoring mechanisms of solidarity at different scales (individual and collective) and in different domains (food production and supply, schooling, accommodation, training, etc.). The cohesion of a society is built on the basis of a feeling of belonging to the same group, in which the fate of each individual is somewhat related to the fate of any other individual. In itself, it is a key element to the achievement of shared identity and in time, to the acknowledgment of a common territory. Social cohesion can come under different forms depending on the context: for instance from insurance mechanisms for industrial countries to food sharing strategies in traditional systems.

Whatever its form, social cohesion plays an important role during crises as it encourages collective response strategies as opposed to individual ones, which can limit the scale of the disaster and its consequences over time.

Social cohesion also influences the possibility of implementing anticipation measures which rely on the willingness of the population to provide immediate efforts for benefits that will result later. This is a crucial point in the current fight against climate change, both from the point of view of greenhouse gas emissions and of the implementation of adaptation strategies and measures.

6.3.2.4. *The diversification of economic and subsistence activities*

Vulnerability is even more pronounced where the number of different activities practised on the territory is limited and where these

10 In a situation where gaps in wealth are not too significant.

are mostly reliant on one environmental resource. In the context of a mono-activity system, the depletion of the dominant sector through a natural disaster cannot be compensated, even for a short while, by the activities in another sector. In this case, and without the intervention of an external source of help, the territory will be deprived of any economic resources at a time in which these are most needed for the reconstruction of its society.

A system relying on a single activity also has the second disadvantage of offering a limited range of jobs, so that, when a hazard occurs, it is the whole productive force of the territory which is affected by it. The loss of jobs and the lack of new perspectives impact the households' income and undermines their ability to recover from a crisis. This sort of vicious circle can quickly become rooted in the system.

The diversification of activities is not however a guarantee in itself to lower vulnerability. These activities must also rely on the use of different resources. Indeed, if a hazard is responsible for permanently degrading the resource in question, the territory will find itself in a similar situation as with the previous case, i.e. a situation where the production system as a whole is damaged.

6.3.2.5. *The political and institutional organization*

The political and institutional model, and by extension the administrative structure, here relates to the mechanisms responsible for the operation of the territory. It expresses *a priori* the fact that a territory made up of territorial units (districts for example) which are poorly-connected will be more fragile when faced with a hazard than a more 'coherent' environment, i.e. whose functioning also relies on the articulation of networks[11]. The consequence of this fragility is twofold. First, the propensity for the hazard to develop into a proper disaster at the scale of the area affected by it, as this micro-territory

11 This is true if in parallel, the articulation of territorial components does not favor the propagation of the consequences of a hazard from the place where the impact occurred to the entire territory. We are touching here on the ambiguity in the operation of network systems, in which we will not go into detail but which shows the challenge in looking at the vulnerability of territories from a general point of view.

will tend to be relatively isolated when dealing with the hazard. Second, the ability to counteract the domino effects characterizing the disaster and to limit the spatial and temporal scales of these effects will tend to be smaller for the area in question. So, although social cohesion and economic diversification for instance can limit the level of vulnerability, other elements which characterize mechanisms of response to crises and the prevention of natural hazards also play a part. These are on the one hand legal and regulatory frameworks applicable over the whole territory and, on the other hand, tools and bodies which help in their successful implementation. These elements can be traditional or take more modern forms.

Two main questions arise. First, do political and institutional bodies (traditional or modern) complement each other in terms of managing and preventing natural hazards over the scale of the whole territory? In other words, are all the domains affected by the risk, and indirectly, all the factors affecting the level of vulnerability, represented? The answer to this question leads to the second question: are these bodies well coordinated? Do they complement each other and, beyond this, are they likely to respond together to the global factors of risk? These questions are particularly relevant in the case of continuous risks, as anticipation mechanisms – and hence long-term regulation – require coherence in political and institutional measures so as to build well structured responses likely to address as soon as possible the domino effects inherent to any hazard.

6.3.2.6. *The living conditions of the population*

The level of development contributes, to the same extent as other factors, to creating/conserving a certain degree of leeway. In this case, we consider living conditions as a reflection of the level of development. Several elements need to be taken into account, including demography, habitat, education, employment status, transport system, energy and health.

The population growth rate, the population densities varying from one place to another and the particular structure of the age pyramid are all factors that influence vulnerability. A strong population rise for instance requires considerable investment in terms of spatial planning (accommodation, telecommunication, public infrastructure, etc.), which can paradoxically impinge on the budgets allocated to the management and response to hazards as well as amplifying its impacts. As an example, the gradual increase in population density on capital islands of coral reef archipelagos is clearly leading to an increase in population numbers exposed to marine-meteorological hazards [DUV 13] .

Living conditions directly affect the vulnerability of local communities in the face of risk. Buildings can be precarious and fail to resist the movements of water, violent winds, high temperatures or even salt intrusions. They can be more or less adapted to changes in climatic conditions.

The vulnerability/education relationship is very ambiguous. On the face of it, it would be reasonable to think that the level of education of the concerned population would tend to condition the capacity of its individuals to adopt behaviors and practices that reduce vulnerability. However, a high level of education does not necessarily include a precise knowledge of all the different types of risk affecting the place or the adequate response mechanisms to deploy. Similarly, traditional knowledge, which is a fully fledged educational mechanism, can instigate adequate anticipation and resilience measures when faced with known hazards. So education should be perceived in the very broad meaning of the term and as a function of the specificities of the local context.

Employment-related aspects which are both quantitative and qualitative (proportion of active population, types of activities, level of wage, informal subsistence activities, etc.), also affect vulnerability in that they determine the level of income for the households. The income is largely responsible for the presence of solid accommodation (either modern or traditional), for transport, specific resources, etc. Depending on the context, income can also explain settling in less risky areas, although this parameter should be analyzed with caution as coastal plains surrounding lagoons, such as the Ermitage in Reunion Island for instance, are highly attractive places where land is associated with very high prices.

Public and private transport, as well as the quality and diversity of the transport network enable managing crises more or less effectively, which can for instance help when evacuating a location or when delivering food and first aid. Thus networks and transport systems can support spatial organization which is adapted to natural and gradual changes, i.e. the re-development of existing or new activities in other parts of the territory. This leads to issues of population access to vehicles, to practicable roads and to fallback spaces where they can be taken care of (hospitals, temporary accommodation, etc.).

The type of energy used by households is a limiting factor or a factor aggravating vulnerability as it reflects the dependence on a source of energy, which is itself more or less sensitive to hazards. Many households in the world use mainly wood to warm their homes or to cook. In the event of fire, floods or in the longer term, reduction of the local production of wood, what effects can be expected from the challenges in harvesting this source of energy? Similarly, a city without electricity over several days can face several problems, whose extent will depend upon the ability of the territory to restore electricity networks and energy production infrastructures.

Finally, sanitary conditions can increase or attenuate the level of vulnerability. In theory, a healthy population tends to be less affected by a hazard and its indirect consequences (mainly poor sanitation). As the healthcare system will have fewer social and health emergencies to attend to, it will in theory have more time to develop long-term and more global adaptation strategies. Here, as for other elements, the marginalization and inequality issues account for the fact that some population groups are more at risk than others. However, it is worth qualifying this statement as many populations which are immune to certain pathologies can be vulnerable to new pathologies. Yet, climate change brings new (or returning) diseases in areas where they had been eradicated or had never been encountered.

The main conclusion that can be drawn at this stage is that, given the diversity of influencing factors, they can only be studied with their potential inter-relations in order to hold a coherent and realistic discourse about vulnerability for a given territory or system. The study of vulnerability is therefore transdisciplinary by nature, where social sciences play a major role.

6.4. From vulnerability to adaptation: theoretical framework

6.4.1. *The dynamic relationships between vulnerability and adaptation*

The essence of the analysis of the vulnerability of a system is to understand how the latter works (what risks threaten it? What challenges? What regulations are required?), and then to imagine future scenarios and propose pragmatic responses to these [DOW 03]. In turn, the implementation of adaptation strategies should aim to reduce the vulnerability of a given territory to natural hazards over the long-term. According to the IPCC still, adaptation is "the adjustment of natural or human systems in response to present and future climatic stimuli or to their effects, in order to mitigate the damage or to exploit beneficial opportunities".

From this point, schools of thought diverge and consider vulnerability as a function of societies' ability to adapt, or conversely, consider adaptation as a function of the level of vulnerability. A different position consists of claiming that two stages of vulnerability can be distinguished depending on whether the climate change dimension is taken into account or not. In addition to climate change concern, every territory is subject to some types of natural hazards. In fact, although climate change could generate some hazards that are unknown to the territory in question, it is most likely to exacerbate its level of exposure to known hazards. Each territory therefore reports a level of vulnerability that is intrinsic (or original) and which is modified by climate change. The resulting vulnerability is a second state which, more than the previous one, will depend on the forms of adaptation implemented by the society (in order to reduce the level of intrinsic vulnerability). Therefore, we claim here that vulnerability is a good indicator in the case of climate change of the efficiency of adaptation measures.

So vulnerability should be understood as a dynamic characteristic which evolves over time (Figure 6.6), and which expands or contracts depending on changes in the context of hazards on the one hand, and on the measures taken by the societies on the other. Virtuous or vicious circles can arise between vulnerability and adaptation and therefore, adaptation should also be considered as dynamic and evolving.

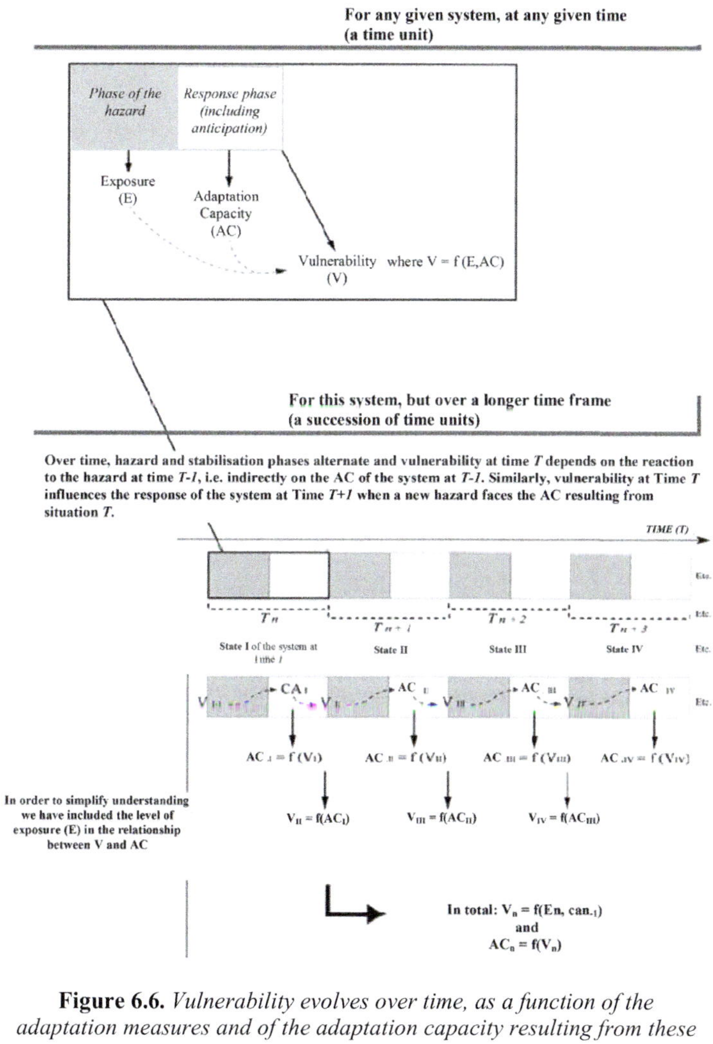

Figure 6.6. *Vulnerability evolves over time, as a function of the adaptation measures and of the adaptation capacity resulting from these*

6.4.2. *Adaptation, a three-dimensional concept (process, state and strategy)*

Adaptation is related to the measures deployed by a society as the context in which it lives changes. In parallel, adaptation is found under specific forms which are expressed as projects, programs, plans and policies. Several dimensions must thus be distinguished to characterize the concept of adaptation exhaustively. So we can see it simultaneously as a *process*, a *state* and a *strategy* [MAG 09]. Adaptation as a *process* expresses the idea of being or not in the process of adapting, adaptation as a *state* refers to being adapted at any point in time, and adaptation as a *strategy* reflects the willingness to adapt (Table 6.2). Differentiating these three dimensions enables the development of scientific studies that are useful to the various types of stakeholders, in terms of raising awareness on the challenges of adaptation to climate change, and of identifying adaptation options that are both realistic (from an implementation point of view) and ambitious by their objectives.

Adaptation dimension	Expresses the fact that...	Synonym	Underlying approach
Process	*Being in the process of adapting*	*Mechanism*	*
State	*Being adapted*	*Form*	*Action*
Strategy	*Being willing to adapt*	*Policy*	*Intention*

* In our opinion there is no voluntary approach as such behind the idea of *process*, as is the case for the *state* and *strategy* dimensions. Here, the dimension of *process* refers to the nature of the system itself and to the ability its various components and their interactions confer to it. In this light, the will to adapt is indirectly reflected within the dimension of *process*, whereas it is directly embodied in the dimensions of *state* and *strategy*.

Table 6.2. *The three dimensions of adaptation: process, state and strategy*

6.4.2.1. *Adaptation as a process*

The adaptation *process* refers to adaptation mechanisms, i.e. to the logic of the system that can account for changes in the various forms of adaptation (projects, programs, plans and policies). As vulnerability and therefore adaptation are necessarily global and systemic, the process dimension integrates both the environmental and human components of the territory under study (see previous influencing factors). This dimension therefore specifically refers to the idea of evolution of the development models to be promoted in order for them to be least vulnerable to natural hazards.

6.4.2.2. *Adaptation as state*

Adaptation as a *state* refers to the practical forms of adaptation on the ground, in the various spheres and at different spatial scales. These forms can be realized in practical initiatives (project, associations, national adaptation plan, change in practices, etc.) or still be at the stage of idea (new ideas for example). The state of adaptation expresses whether or not a society is adapted to its natural environment at any given point in time. This refers to a certain level of unity in time and therefore to a specific timescale. In other words, whereas it can be said that a society is adapted (or not) to the current variability in climate, it appears at first glance impossible to say whether the same society is adapted (or not) to climate change. Indeed climate change expresses conditions which are necessarily changing and therefore which cannot exactly be predicted in advance. So a society which is adapted to the climatic conditions at any given time T could be maladapted anymore at time T_{+1}, and vice versa. Let us note here that the discussion around relevant timescales refers to the idea of evolving adaptation mechanisms and hence to the *process* dimension. It appears that the first two dimensions, as well as the next one, complement each other, which contributes to the coherence of the reading framework proposed in the present study.

6.4.2.3. *Adaptation as a strategy*

Finally, the third dimension sees in adaptation a *strategy*, i.e. a policy. Adaptation here no longer refers to a logic of *process* or action in the strictest meaning of the term (*state*), but rather to an intention to adapt through the ways of occupying territories, of governing societies, of planning development, etc. There is both an idea of intention to adapt and the idea of being conscious of the present and future challenges. There again, the changing nature of adaptation is put forward and, as opposed to the dimension of *state*, adaptation as a strategy assumes a mixture of different timescales, and it also involves taking into account the various spatial scales.

6.4.3. *Thinking in terms of pathways toward adaptation*

This form of categorization may appear somewhat artificial as the dimensions of *process*, *state* and *strategy* are interrelated through causality relationships and they come together more than they differ. However, this distinction is useful, as in addition to encouraging conceptual thinking, the exercise can favor the tailoring of scientific knowledge around adaptation to the discourse of different types of stakeholders whose culture and filters of analysis may be different. Not everyone of them speaks the same language, although they all talk about the same thing, i.e. implementing adaptation. Theoretically, national political leaders will for instance be more interested in the *strategy* entry angle (what political choices should be made for what medium- to long-term vision?). Tourism operators facing problems of coastal erosion for example will look for solutions on a shorter term, and will therefore look at adaptation as a *state*. As to anthropologists or sociologists for instance, they will probably be more inclined to looking at the *process* dimension of adaptation. So we support the idea according to which these three dimensions can help in improving the integration of scientific knowledge within complex social dynamics, including at different territorial scales. It can help bringing stakeholders together around a similar language. Elaborating adaptation strategies that are relevant because they are adapted to the context, can be greatly beneficial within this three-dimensional analysis framework. The latter enables through segmentation, on the one hand to identify all the components of a territory/society which are involved in adaptation, and on the other hand, to make different pathways emerge with different types of tools (political, regulatory, social, etc.) so as to persue adaptation. The challenge consists of establishing a link between these components and the stakeholders, which involves gathering them around the notion of *adaptation trajectory*.

By *adaptation trajectory* we mean the pathway through which a territory attempts to adapt to climate variability and change. The notion of trajectory expresses the idea that adaptation is necessarily dynamic. Adaptation should not be understood as the quest for a still image in the future ('being adapted in 2050 or in 2100'), but rather as a state of relative balance between humans and their environment which will never be fixed forever but rather which will constantly require more or less profound readjustments. In other words, a great challenge consists of not only attempting to determine today the future forms of adaptation (*state* dimension), but also identifying the broad directions of adaptation (*strategy* dimension) which will enable maintaining a certain level of flexibility from the point of view of implementation (*process* dimension). This refers to the idea of *flexibility*: given the uncertainties around the local impacts of climate change, it seems obvious to us that building robust adaptation strategies must rely on the one hand on the good understanding and management of the adaptation process, and on the other hand on the creation of reference visions for the future ('ideal' state of adaptation). The notion of trajectory establishes the link between the two as it aims to determine how to reach these reference points. So the *adaptation trajectory* is necessarily a cross between the adaptation *process*, the successive adaptation *states* and the adaptation *strategies*.

6.5. The action framework: thinking of adaptation as being at the crossroads of anticipation and resilience

To summarize what has been said so far, and from a very general point of view, it can be argued that the essence of adaptation is to allow societies to reduce their degree of vulnerability to environmental changes, both current and future, sudden and gradual. So it can only be seen as dynamic and following three dimensions, as it is a *process*, a *state* and a *strategy*. What remains to be discussed is how this theoretical framework can turn into a framework for action.

The theoretical approach can be expressed as follows: the challenge of adapting to climate change consists of making current choices compatible with future environmental challenges. This is also one of the principles at the basis of sustainable development with the particularity in the context of climate change that we should start adapting now to future environmental changes which, at local scales, remain tainted with considerable uncertainties. In order to go beyond the problem of uncertainty, which cannot possibly be used as an excuse for the lack of action, an interesting option for societies consists of maintaining a maximum of room for manoeuvre so as to always be able to readjust the development choices made as knowledge progresses. In other words, societies not only have to be sufficiently solid to project their development strategies over time but they should also be flexible enough to be able to readjust these strategies. So the notion of flexibility is central and it relies on two approaches to the future which complement each other: anticipation and resilience [MAG 13]. In theory, the first approach refers to the ability of a society to project itself into the future, whereas the second refers to the ability of societies to increase their response capability when faced with relatively unpredictable hazards.

6.5.1. *Anticipating for adaptation*

Anticipation is the most common dimension of adaptation to climate change. It refers to the long term and to the ability of imagining and initiating responses ahead of the problems. The time horizons of 2030, 2050 and even 2100 are often evoked as they correspond to timesteps in climate model simulations, and as they are deadlines for which it is thought that it is necessary and desirable to think about fighting climate change impacts. Moreover, any society needs time to change its consumption patterns, its spatial planning, its relation to the environment, etc. Making the way our systems function evolve in a few decades involves setting changes in motion now, or in other words, anticipating.

One of the main questions raised is that of the ability of political leaders, economic stakeholders, local organizations and populations to address uncertainty around future climate change impacts. How to prepare for something that cannot be clearly identified? The question raises many challenges but we must also remember that climate change-related uncertainty is only part of the problem. Indeed, climate change will act in parallel with other drivers of change: economic trends, changes in geo-political relations, fashion phenomena, other natural hazards (earthquakes, tsunamis, volcanic eruptions, etc.), the widening of socio-economic inequalities, etc. However, the latter are also surrounded by uncertainties. So uncertainties associated to climate simulations, before bringing up new questions, first reinforce a challenge that already exists, i.e. taking the chance on future benefits that can be gained by acting now. Thus anticipation is not a new challenge, although the reality of climate change makes it now a vital necessity.

Furthermore, the fact that there are uncertainties around the local impacts of climate change for the next decades does not mean that future hazards will necessarily be new. On the contrary, and at least for coastal areas, a number of threats brought about by climate change will mainly exacerbate the natural hazards that in some cases have long been hitting many territories. So there is already a certain level of experience of past risks which can be put to use to better anticipate future risks [DUV 14]. And as it cannot be excluded that some territories be confronted in the future by hazards they had never or not frequently encountered before, they could at least in part benefit from the experience of other territories. This realization enables us to take a step back on the issue of uncertainty and to transmit a positive message, namely that the agents of adaptation are not entirely clueless in the face of climate change. On the contrary, they are already empowered with means of intervention.

Ex-ante initiatives cannot be sufficient in themselves as hazards cannot be expected to be entirely foreseeable. A complementary approach should therefore be initiated in parallel and it should consist in favoring the ability of societies to respond to unpredictable natural hazards.

6.5.2. *Remaining or becoming resilient*

Without going into too many terminology details of a concept that has been studied at least since the 1970s, resilience can be defined as the mechanism that enables a system to find some sort of balance after having been victim of a hazard[12]. This mechanism is therefore involved *ex-post* as opposed to anticipation. However it also

12 Whether the balance is that present prior to the hazard or whether it is a different type of balance which integrates the changes initiated by the hazard. This question remains a hot topic within the scientific community.

intervenes in the process of adaptation. Indeed developing or maintaining the ability to respond to an unpredictable event remains a topical challenge as climate variability, which accounts for the occurrence of extreme events, will necessarily act in combination with climate change as such[13] [BER 07]. So reinforcing our ability to 'digest' crises that will never cease to occur is a key-objective of the resilience pillar. There again, our current experience can be useful as natural hazards which are very similar to future hazards are already taking place. So the adaptation challenge consists of projecting the capacity or lack thereof of a territory to be resilient in the future.

Scientific studies are already teaching us that resilience relies principally on the consolidation of spatial coherence and social cohesion, on the improvement of living conditions (accommodation, health, food, etc.) and more generally on the conditions of development (education, public policies, employment, etc.), and finally on a better preservation/conservation of natural environments and their biodiversity (ecosystem services among others). This brings us back to the influencing factors of vulnerability previously described and, here again, the aim is to solidify the basic characteristics of territories without freezing them. The flexibility imperative re-emerges: as the distant future cannot be predicted with certainty, a promising strategy consists of being capable of responding "in real time"; yet, a certain level of flexibility in the operation of a system affects its way to respond, and in time, its ability to show resilience.

6.5.3. *The overlap between anticipation and resilience*

As the mechanism of resilience consists of the response to a hazard and as it is part of a short- to medium-term dimension, resilience is a reaction to the gaps in the anticipation mechanism. The latter indeed does not act in response to a hazard and is established over medium to long timescales. Conversely, the anticipation logic fills the gaps in the

---

13 Globally, climate change characterizes the evolution of the climate's parametric means (at the scale of several decades), whereas climate variability characterize changes in these parameters around a mean. Both phenomena can therefore be combined.

resilience mechanism, and for this reason, these are the two complementary pillars behind the flexibility of societies and their territories, and therefore of their ability to adapt to future climate change.

It can be deduced that the long term is not the only time horizon to be considered in terms of adaptation, and that the short time, that of practical action, is just as important. In other words, the adaptation strategies cannot dissociate themselves from the overlap in timescales, so that actions today which are associated with immediate benefits are not disconnected from the challenges of adaptation to climate change. Therefore, many actions favoring adaptation (e.g. revising spatial planning through the integration of risks, or the fighting of poverty and social inequalities) can already be undertaken, on the condition of course that they are understood with the objective of reinforcing both the resilience of a system and its anticipation capacity.

6.6. Conclusion

This study proposes an approach to the vulnerability and the adaptation to climate change concepts which enables a wide range of researchers in social sciences (from different disciplines, with different levels of scientific maturity) to grasp these and contribute, through fieldwork, to improving knowledge.

The study of vulnerability to climate change, and even more so of adaptation to climate change, arose relatively recently with the realization of strong interactions between human societies and natural phenomena. Although so-called hard sciences (climate science, geomorphology, oceanography, etc.) play a crucial role in defining the nature, the intensity and the frequency of future natural hazards and environmental changes, the same applies to social sciences. The responsibility of social sciences will be to shed light on the human dimensions of vulnerability and risk. Various dimensions (environmental, cultural, political, etc.) and timescales should therefore be combined in order to achieve this and it is this overlapping of dimensions and scales which sets the foundations for the legitimacy for the concept of vulnerability.

This study provides a framework of analysis which relies on six broad factors affecting vulnerability to climate change and, more broadly, to global change: spatial configuration, ecosystem sensitivity, social cohesion, diversification of economic and subsistence activities, political and institutional organization, living conditions. Looking through the lens of these six factors enables us to understand the context in which the risk occurs (components and processes in action), which is useful to the elaboration of a response strategy. So from vulnerability, we can move on to adaptation, whose aim is to reduce vulnerability in the long term.

The text also proposes a framework of analysis for adaptation to climate change. It puts forward, among others, the idea that adaptation should be looked at both as a *process*, a *state* and a *strategy*. Adaptation as a process reflects whether or not the society is in the process of adapting. Adaptation as a state expresses whether or not a society is adapted at any point in time. Finally, adaptation as a strategy refers to the desire or lack thereof to adapt. As these three dimensions overlap, they help in identifying comprehensive, dynamic and flexible ways to fight climate change impacts. The notion of *adaptation trajectory* therefore explicitly refers to the dynamic nature both of vulnerability and of adaptation.

Finally, the study provides insights on how to move from a theoretical approach of adaptation to a more practical action framework. The theory can be summarized as follows: the challenge with adaptation to climate change consists of making current choices compatible with future environmental challenges. As climate change modeling is not, and will never be, a tool which sufficiently predicts, in the long term, all the environmental changes, there is and always will be a part – a part only – of uncertainty around future hazards, including in terms of extreme events. This realization compels us to place at the heart of a society's ability to adapt to climate change, the notion of *flexibility*. Yet, two of the central pillars of flexibility are *anticipation* and *resilience*. Anticipation refers to the ability of a society to project itself over time, whereas resilience refers to its ability in increasing its capacity to respond to unexpected hazards.

6.7. Bibliography

[ADG 06] ADGER W.N., "Vulnerability", *Global Environmental Change*, vol. 16, pp. 268–281, 2006.

[BAN 01] BANKOFF G., "Rendering the World unsafe: "vulnerability" as Western discourse", *Disasters*, vol. 25, no. 1, pp. 19–35, 2001.

[BAN 04] BANKOFF G., FRERKS G., HILHORST D., (eds.), *Mapping Vulnerability: Disasters, Development and People*, Earthscan, London, 2004.

[BEC 01] BECK U., *Risk Society: Towards a New Modernity*, SAGE Publications, London, p. 260, 1992.

[BEC 12] BECERRA S., "Vulnérabilité, risques et environnement: l'itinéraire chaotique d'un paradigme sociologique contemporain", *VertigO*, vol. 12, no. 1, pp. 1–23, 2012.

[BER 07] BERKES F., "Understanding uncertainty and reducing vulnerability: lessons from resilience thinking", *Natural Hazards*, vol. 41, pp. 283–295, 2007.

[BLA 94] BLAIKIE P., CANNON T., DAVIS I., *et al.*, *At Risk: Natural Hazards, People's Vulnerability and Disaster*, Routledge, 1st ed., London, p. 436, 1994.

[BRO 03] BROOKS N., Vulnerability, risk and adaptation: a conceptual framework, Tyndall Center for Climate Change Research, working paper 38, 2003. Available at: www.tyndall.ac.uk/publications/working_papers/wp38.pdf.

[CAR 04] CARDONA O.D., "The need for rethinking the concepts of vulnerability and risk from a holistic perspective: a necessary review and criticism for effective risk management", in BANKOFF G., FRERKS G., HILHORST D., (eds.), *Mapping Vulnerability: Disasters, Development and People*, Earthscan, London, pp. 37–51, 2004.

[COL 99] COLLIGNON B., "Les fondements territoriaux de l'identité inuit d'hier et d'aujourd'hui", in BONNEMAISON J., CANBREZY L., QUINTY-BOURGEOIS L., (eds.), *Les territoires de l'identité – Le territoire, lien ou frontière?*, L'Harmattan, Paris, vol. 1, pp. 93–109, 1999.

[COS 05] COSAERT P., "Les pays touchés par le tsunami: un contexte économique, politique et humain très diversifié", *Les Cahiers d'Outre-mer*, vol. 229, pp. 89–95, 2005.

[DAU 01] DAUPHINE A., *Risques et Catastrophes: observer, spatialiser, comprendre, gérer*, A. Colin, Paris, p. 288, 2001.

[DAU 07] DAUPHINE A., PROVITOLO D., "La résilience: un concept pour la gestion des risques", *Annales de Géographie*, vol. 654, pp. 115–124, 2007.

[DIA 97] DIAMOND J., *Guns, Germs and Steel: The Fates of Human Societies*, W.W. Norton, New York, p. 480, 1997.

[DIA 05] DIAMOND J., *Collapse: How Societies Choose to Fail or Succeed*, Viking Press, New York, p. 575, 2005.

[DIM 91] DI MEO G., "De l'espace subjectif à l'espace objectif : l'itinéraire d'un labyrinthe", *L'espace géographique*, vol. 2, pp. 359–373, 1991.

[DOW 03] DOWNING T.E., PATWARDHAN A., Vulnerability Assessment for Climate Adaptation, APF Technical Paper 3, UNDP, New York City, 2003. Available at: www.undp.org/gef/05/documents/publications/apf-technical-paper03.pdf.

[DUV 12] DUVAT V., MAGNAN A., *Ces îles qui pourraient disparaître*, Editions Le Pommier-Belin, Paris, p. 191, 2012.

[DUV 13] DUVAT V., MAGNAN A., POUGET F., "Exposure of atoll population to coastal erosion and flooding: a South tarawa assessment, Kiribati", *Sustainability Science*, vol. 8, no. 3, pp. 423–440, 2013.

[DUV 14] DUVAT V., MAGNAN A., *Des catastrophes... "naturelles"?*, Editions Le Pommier-Belin, Paris, p. 311, 2014.

[FRA 03] FRASER E.D.G., MABEE W., SLAYMAKER O., "Mutual vulnerability, mutual dependance. The reflexive relation between human society and the environment", *Global Environmental Change*, vol. 13, pp. 137–144, 2003.

[FUS 06] FÜSSEL H.M., KLEIN R.J.T., "Climate change vulnerability assessments: an evolution of conceptual thinking", *Climatic Change*, vol. 75, pp. 301–329, 2006.

[HEW 83] HEWITT K., *Interpretation of Calamity From the Viewpoint of Human Ecology*, Allen and Unwin, Boston, p. 304, 1983.

[HIL 04] HILHORST D., "Complexity and diversity: unlocking social domains of disaster response", in BANKOFF G., FRERKS G., HILHORST D., (eds.), *Mapping Vulnerability: Disasters, Development and People*, Earthscan, London, pp. 52–66, 2004.

[HIN 11] HINKEL J., "“Indicators of vulnerability and adaptive capacity": towards a clarification of the science-policy interface", *Global Environmental Change*, vol. 21, pp. 198–208, 2011.

[LAN 05] LANDY F., "Après le tsunami: l'Inde est bien le pays "émergeant"!", *La lettre Intergéo*, pp. 23–27, 2005.

[MAG 06] MAGNAN A., "L'évacuation des populations aux Maldives après le tsunami du 26 December 2004", *Mappemonde*, vol. 84, no. 4, 2006. Available at: http://mappemonde.mgm.fr.

[MAG 09] MAGNAN A., "Proposition d'une trame de recherche pour appréhender la capacité d'adaptation au changement climatique", *VertigO*, vol. 9, no. 3, 2009. Available at: http://vertigo.revues.org/9189.

[MAG 12] MAGNAN A., "Evaluer *ex ante* la pertinence de projets locaux d'adaptation au changement climatique", *VertigO*, vol. 12, no. 3, 2012. Available at: http://vertigo.revues.org/13000.

[MAG 13] MAGNAN A., *Changement climatique: tous vulnérables? Repenser les inégalités*, Editions des Presses de la Rue d'Ulm, Paris, p. 66, 2013.

[OBR 04] O'BRIEN K., ERIKSEN S., SCHJOLDEN A., *et al.*, What's in a word? Conflicting interpretations of vulnerability in climate change research, Center for International Climate and Environmental Research, working paper 4, 2004. Available at: www.cicero.uio.no/media/2681.pdf.

[OUE 04] OUESLATI A., *Littoral et aménagement en Tunisie*, Editions de la FLSH de Tunis, Tunis, p. 256, 2004.

[PAS 85] PASKOFF R., *Les plages de la Tunisie*, Editec, Caen, p. 192, 1985.

[PEG 96] PEGUY C.P., *L'horizontal et le vertical*, Reclus, Montpellier, p. 98, 1996.

[PER 01] PERRY A., "More heat and drought: can mediterranean tourism survive or prosper?", in MATZARAKIS A., DE FREITAG C.R., (eds.), *Proceedings of the 1st International Workshop on Climate, Tourism and Recreation* (Greece, 5–10 October 2001), pp. 35–40, 2001.

[PRO 07] PROVITOLO D., "Un exemple d'effets de dominos: la panique dans les catastrophes urbaines", *Cybergeo*, 2007. Available at: www.cybergeo.eu/index2991.html.

[REG 06] REGHEZZA M., Réflexions autour de la vulnérabilité métropolitaine : la métropole parisienne face au risque de crue centennale (notamment pp. 49–81), Doctoral Thesis in Geography, University of Paris X, 2006.

[VEY 07] VEYRET Y., (ed.), *Dictionnaire de l'Environnement*, A. Colin, Paris, p. 404, 2007.

[WIS 04] WISNER B., "Assessment of capability and vulnerability", in BANKOFF G., FRERKS G., HILHORST D., (eds.), *Mapping Vulnerability: Disasters, Development and People*, Earthscan, London, pp. 183–193, 2004.

7

Anthropological Approach to Vulnerability and Major Hazards

7.1. General introduction

On an anthropological level, the notions of environment and milieu, as well as those of risk and vulnerability, are understood in a sense that sometimes differs widely from their ordinary usages and their meanings in other disciplines. The reason for this divergence is simple. The definitions of these terms used from an anthropological perspective must be in a position to convey realities that sometimes vary so much as to be asymmetrical. As we will see, what is perceived as risk by some individuals can be considered a stroke of good luck by others, and what is seen as a catastrophe by one community of people can transpire to be a desired strategy for another group. In the first instance (section 7.2), there will therefore be the question of defining these terminologies. In the second instance, by means of two case studies, we will study the cultural variation in perceptions of and methods of handling major hazards.

The first case (section 7.3), which concerns one of the largest volcanic eruptions of the 20th Century (at Ambrym in 1913), will allow us to illustrate one of the linchpins for an anthropological understanding of the relations between human societies and factors that threaten their existences. There is no catastrophe, that is to say

---

Chapter written by Yoann MOREAU.

disorder, but only that which is relative to a cosmology, that is to say relative to a historically established order. In other words, human societies evaluate the risks and dangers that could happen to them by the light of their own world view. For reasons of simplicity, our approach will only take account of two major anthropological schemes: first, the so-called "modern" societies that believe that their principle of general organization arises mainly from *phusis* (the laws of matter) and, second, the so-called "traditional" societies that believe that their principle of order arises above all from *nomos* (social rules).

In the conclusion section of this study, we will see, from the second case (section 7.4), that what constitutes a catastrophe is not always entirely negative. Within a single society, some individuals are the victims whereas others profit from what has happened. In other words, at a sociological level, there is always ambivalence toward what happens, "the misfortune of some is the good fortune of others" says the proverb. To go beyond the simple good sense of the proverb, the in-depth study of the media treatment of the earthquake that struck Edo (the old name for Tokyo) in 1855 will enable us to grasp the intrinsic ambivalence associated with the reception of major hazards. Finally, in section 7.5, an analysis will be presented, which will enable us, on the one hand, to consider the profoundly ambivalent character of all major hazards and, on the other hand, to demonstrate its complexity in a synoptic form. This is in order to avoid succumbing to a common tendency in the interpretation of catastrophes: catastrophism.

7.2. Definitions: "environment and milieu" and "risk and danger"

In the following, the word "environment" designates objective reality, that is to say what every individual must confront in order to exist. In other words, whatever its form of life, every individual lives within an environment that is shared with the entire living world. Whether one is Papua or Breton, human or dog, or tick or lichen, the environment is the same for all. However, although we live in a shared environment, we perceive it in different ways. A Papuan does not pay attention to the same thing as a Breton, a dog is not sensitive to the same stimuli as a human, and similarly a tick and a lichen. Each behaves depending on what he/she perceives, that is to say not depending on the environment "in itself" but on the aspects that appear significant or vital to him/her. Perception of the environment varies not only depending on innate qualities (belonging to a species), but also according to the acquired aptitudes (belonging to a developing milieu). The perceptive matrix is not the same from one individual to another; it depends on species and on fields of experience that have been developed during its growth.

We are indebted to Baron Jacob Von Uexküll who demonstrated [UEX 56] that each species does not act according to the environment (*Umgebung*) but according to its milieu (*Umwelt*). This means that a form of life, whatever it may be, cannot perceive the environment, but only its organic and semiotic manifestation. The environment is "veiled" from an inate organic part (which defines a particular field of organic stimulations) and an acquired cultural part (which favors certain types of sensitivities and experiences). In other words, and this is the definition that we propose, a milieu is the concrete dimension of the environment, something with which a life form interacts in a concrete sense. Von Uexküll was able to demonstrate this on an ethological level, but he only sketched the specificity of human milieux. Indeed, these milieux are characterized by a very strong salience of the cultural dimension. Humans, more than any other species, acquire and share competences that they do not have at birth.

We are indebted to the paleoanthropologist André Leroi-Gourhan for demonstrating the crucial importance of the "social part" in the formation of human milieux. His main thesis, the externalizing of symbolic dimensions in technical systems over the lengthy time-span of biological evolution [LER 64], shows that the evolution of the species *Homo sapiens* is linked to a structuring of environmental transformations in a human milieu. The interrelationship between *Sapiens* and their surroundings has become consequential as it has been systemized. Unlike other species, humans have organized their meanings in languages and their actions into techniques.

From *Homo habilis* to *H. sapiens*, generations of hominids have progressively used their milieu in an increasingly ample and structured technical, symbolic, energetic and ecological system. This amounts to a "triple and mutual creation" [BER 00, p. 96] of the human species in relation to that surrounds it:

– a hominization (the species becomes *sapiens* by exteriorizing its actions in symbols);

– a humanization (*sapiens* diversify culturally by organizing their symbols into languages);

– an anthropization (the environment conforms increasingly progressively to human needs).

This process of "hominescence" [MOR 13] explains why the human milieu very particularly modifies the environment (that is to say, the milieu shared by all species). By adapting what surrounds them to their own needs, humans do what other life forms do but, because of the organization of symbols and techniques into systems (of languages and technocraties), the transformations' orders of size have been considerably multiplied.

A very strong interdependance has been built up between humans and their environments. On the one hand, the scale of human impact on their surroundings is taken to a geological level. This is the anthropocene hypothesis [STE 11] according to which the environmental conditions that affect all life forms (that is to say the biosphere in its entirety) are profoundly linked to human behavior [BON 13]. On the other hand, contemporary humans would not know how to live without the technological (from clothes to habitations, via tools and the use of non-edible energy resources) and symbolic modifications (from languages to information systems, from moral values to ethical concerns, etc.) through the means of which they are in a position to create and interact with what surrounds them. Due to the work of Leroi-Gourhan, we know that this coevolution and codependance is in continuity with a long-standing historic trajectory.

It fell to the geographer Augustin Berque to show that not only does the human species exteriorize its animal functions in tools (it makes knives, for example, which assume – among others – the function of the claws and teeth of carnivores), but also it equally operates the reverse: it interiorizes its tools with a symbolic value (a knife has connotations of a phallic dimension, for example). These movements of coming and going – exteriorization (technical systems) and interiorization (symbolic systems) – characterize the human relationship to its geographical and social milieu [WAT 11]. This double movement of objectivation (of the milieu into an environment) and subjectivation (from the milieu into symbolic systems) conveys the manner in which a human milieu is structured and deployed. This alternance of the process of objectivation and subjectivation is called "trajection" or "mediance" [BER 90]. The first term emphasizes the rythmic dimension (alternance of opposing and complementary processes, for example subjectivation and objectivation); the second term emphasizes the exhibition and organization of this relationship that concerns neither the objective environment nor a subjective representation, but a milieu.

From this perspective, the evaluation of the impact of a major hazard can only happen in relation to the way in which a milieu is inhabited and invested. It is therefore a question, as far as possible, of taking account of the maximum of the technical, symbolic and ecological dimensions specific to the human milieux touched by a catastrophe. Given that these are extremely diverse, it is useful to proceed by means of a conceptual aparatus which makes this diversity accessible and possible to study.

It fell to Philippe Descola to propose a "grammar of cosmologies" [DES 05], which enables us to grasp the profound logic of the organization of human milieux. The anthropologist distinguishes four schemes of transformation from environment to human milieu. There would thus be societies that were in the main "naturalist", others mainly "animist", "totemist" or "analogist". At each of these trajectories of "making a milieu" correspond specific vulnerabilities and ways of "deconstructing a world" [MOR 13]. From the "naturalist" point of view, the hazards are mainly evaluated in terms of material causes (which act on flesh and materials); for "animist" communities, they are estimated in terms of the motives (i.e. what motivates an action) of responsible agents; for "totemist" groups, they are assessed according to a historic reason (which explains the contingence of an event) stemming from a genealogical framework (that of totems); for "analogist" societies, they are analyzed according to a principle (on which rules and the rule of law are founded) in the light of which the forming of an imbalance is perceived. Without entering here into the details of these categories (i.e. causes, motives, reasons and principles), it is sufficient for us to know that they exist, that is to remember that the evaluation of the impact of a hazard is linked to the cultural trajectories that distinguish specific human milieu from a common environment.

Thus, the analysis of hazards in terms of "risks", that is to say in terms of the potential for harm from a material point of view (death and physical damange), arises from a specifically naturalist perspective. In this regard, if we speak of civilization or soceties "at risk" [BEC 01, GID 94, LAG 81], it is from a particular human milieu (modernity) whose relationship with its surrounding is based on a materialistic framework. Correlatively, from the point of view of these societies, the evaluation of dangers, that is to say the potential for immaterial harm (fears, anxieties, depressions, etc.), appears, overall, to be underestimated [DOU 82]. Conversely, societies whose principal schema of explanation is of the animist type (evaluation of the impact in terms of moral and symbolic responsibilities) will have a tendency to underestimate the risks and overestimate the dangers [MOR 08].

Let us simply retain these two definitions:

– a risk is a material hazard, which threatens the bodily existence of individuals (blows, injuries and physiological deaths) and can cause tangible and quantifiable damage (destruction of materials or of economic components);

– a danger is an immaterial hazard, which threatens the moral well-being of individuals (fears, anxieties and psycological depressions) and can generate intangible but qualifiable impacts (acculturation of a symbolic and moral order, loss of traditions and customs, breakdown – indeed rupture – of social consensus, etc.).

Let us resume that in order to evaluate the impact of a hazard, it is wise to pay attention to the milieux (that is to say to the ways in which the environment is concretely lived and perceived) and mediances (that is to say the ways in which the milieux are organized and structured). Depending on the societies and the domains of expertise, attention is diverted as a priority to material or symbolic hazards (risks or dangers), to the blows struck at the continuity of meaning (an event) or on the deregulation phenomena (process). We will have recourse to these categories in the two case studies that follow.

7.3. Ambrym 1913

In the autumn of 1892, Robert Lamb, a doctor who qualified at Edinburgh and had just been ordained a priest by the clergy at Auckland was sent by the Presbyterian church to establish a religious, medical and industrial center at Ambrym. The island communities then consisted only of 5,000 natives known for their intensive use of magic and *abelyep* sorcery. Previous missionnaries (1862, 1883 and 1885) named the island "the Mother of obscurity". Lamb saw a challenge there that was even greater. It was a question, he noted, of "succouring [and] illuminating [these] dusty peoples" [LAM 05]. Medical care is only one dimension of a sanitary enterprise of a biological, moral and economic nature where it is a question of civilizing "savagery" in all its forms.

The reverend made his home in the west of the island, at Dip Point, "exactly along the passage of the boats and with ten surrounding villages within range of gunshot" [LAM 05, pp. 43–44]. He thus planned to enlighten the whole of the archipelago. However, only a few months after his arrival, in March 1893, a hurricane destroyed the crops and forests, as well as the buildings that Doctor Lamb, his wife Mary Manson Reiach and his companions had established. A famine followed. Their twins, born there three months previously, suffered from malnutrition and then succumbed two weeks later to an intense fever.

A year later, Lamb rebuilt the hospital. It burned down the same year. Having no more funds, he returned to New Zealand to enquire at the Presbyterian church, returned in 1896 and refounded a hospital in the same area. That year, it succeeded. He was pleased since, he noted, 26 patients went there each day on average. The following year, however, Doctor Lamb contracted tuberculosis, which eventually forced him to leave Ambrym and the New Hebrides. This series of misfortunes in succession added Lamb to the community of preachers who named the island the "Mother of Obscurity".

Seventeen years later, in the spring of 1913, Dip Point was provided with a hospital, a church and a school that were active under the auspices of a new administrator, Doctor Bowie. However, on the 7th December, a joint eruption of the two active volcanos on Ambrym, Benbow and Marum, caused all these edifices to disappear and all the settlers to flee. According to Doctor Bowie's testimony, "close to Dip Point six distinct craters surfaced" [MAR 14, p. 390], torrents of water and lava flowed there for some days. The hospital in particular was destroyed and swallowed up by the formation of a new crater, and its site as well. The island was unrecognizable, some land had disappeared, other land was formed, entire forests and villages were burned, ashes and dead animals were strewn across the territory. The mission to Ambrym would remain in the annals of the Presbyterian church as having been the most costly in terms of human resource and finance [MIL 89, p. 10].

7.3.1. *The point of view of the settlers*

For historians and geologists, this eruption was the most devastating of the 20th Century in the New Hebrides. For the non-native inhabitants of the island – Robert Fletcher (a graduate of Oxford, having fled to the New Hebrides from a professorship that he hated), the Reverend Maurice Frater and Doctor Bowie – the volcanic phenomenon was initially sublime (an ambiguous idea in itself). For Doctor Bowie, whose observations were noted by Professor Marshall:

> The view was magnificent, everything inspiring awe. The lava, partially incandescent, burnt the enormous trees, throwing them in an embrace into the air. A little after 3 o'clock in the morning the lava sunk with a wild hissing noise sauvage and disappeared roaring powerfully into the sea. The view was then superb, and I will never forget it.

> The din of the lava flowing away across the forest sounded very much like the noise of a hurricane. [MAR 14, p. 389]

Robert Fletcher, the morning of the 5th December, wrote the following to his old friend and colleague Bohun Lynch:

> Yesterday, I spent the evening, sitting on a folding chair, on the shore, to contemplate the most beautiful spectacle a man could wish to witness. The moonlight was splendid. The waves were breaking on the reef with a noise like thunder and forming a broad, white and radiant strip. In the bush, the fire roared and bellowed, the great trees came crashing down with a roar. The ravines bubbled like giant serpents of flame that crawled towards the sea. And I was there to smoke a good pipe and drink a glass of whisky and water from time to time, absolutely incapable of imagining that I was running the least danger. [FLE 89]

This almost "joyous" attitude, initially blind to the destructive potential of the volcanoes, was not shared by the natives who worked for Fletcher. He had to threaten them with his winchester, "otherwise they would have fled into the sea". What Fletcher feared, in fact, at that moment was not the eruption – of which he enjoyed the spectacle – but the thought of finding himself alone face to face with the locals who were "not easy to handle ".

This initial subjugation rapidly gave way to worry and terror unusual in the old adventurer from Oxford. The following day, he didn't write. His next letter, dated the 7 December, relates the brutal degradation of the "spectacle " very briefly:

> This letter, I am beginning, at least, in bizarre circumstances. Pay good attention. I expect every hour (even every minute or from one second to another) to be engulfed by a flow of lava thrown up by a volcano, with pestilence and death. [FLE 89, p. 89]

On 12 December, while the main volcanic activity had stopped on the 10th, he again took time to write. He recounted his flight, having crossed boiling waters to reach the cliffs:

> There was no safe place. My boys and I could do more from tiredness. We had been made up by groups of natives from the bush fleeing in all directions, naked and mad with terror. The seashore was covered with a foot of ash, mixed with dead and boiled fish, with tortoises, birds with burned plumage, and others rejoining paths of flight. [FLE 89, p. 91]

To summarize, for the settlers, missions of religious conversion and medical hygiene form part of a civilizing step toward education and the saving of "dusty" peoples. When the volcanic eruption began, while the indigenous peoples fled to all parts, they were staggered by the "spectacle" of sublime scenes "magnicent, all inspiring awe". On a second occasion, but only when the lava and ash were at their feet, were they horrifed. They then experienced and described the eruption as catastrophic. Remember these two occasions: a metaphorical blindness resulting from amazement followed by catastrophe, and this double perception, the positivity of the conversion enterprise and the radical negativity of the volcanic eruption.

7.3.2. *The native point of view*

It was an internal conflict that, in the words of the Ambrymais, caused the arrival of the settlers. Disputes between families had undermined social relations for several years, and the settlers were welcomed in this context in the hope that they could play the role of intercessors. Harold Temar recounts this as follows:

> Before, in the time when the elders were present, there was a conflict; there were disputes between. [The elders] therefore invited the missionnaries so that they could restore peace. [HOS 07]

The Ambrymais were delighted by the arrival of the newcomers. But they became progressively disenchanted when they saw what the converts became following religious conversion or education. The loss of ancestral traditions and the progressive obsolescence of traditional customs worried them to the highest degree. Trees were felled and the sacred fires were violated. The Ancients were angry, they saw the rules of community life eroded as the Presbyterian school filled with more and more recruits. Harold Temar recounts – and this is a version that is the consensus today among the Ambrymais population:

> The pupils left for anywhere and broke the laws of our customs. The Elders said: Christianity has arrived and we are going to lose our customs! They said to themselves that they would all lose their customs. [HOS 07]

The conversion enterprise experienced, indeed, a high point at the beginning of 1913. The reports of the doctors from the hospital also make mention of an important growth in the rythm of the mission. The process of acculturation accelerated, but this did not arise from community decision-making. According to the accounts of many Ambrymais (including Harold), it was a large assembly organized by Doctor Bowie that would unite all the locals, pushing the elders and other members of the population to react:

> This happened on 28th November 1913. On this day, the missionnaries made an enoumous meal. All the main chiefs were there, the Mél, the Naim Meleun, those who had abandoned the sacred fire and become Cristians. [...] Every one together. When the Big Men of the bush received the news, they said: "no, this school has arrived and is affecting our customs, it will make lose our customs". [HOS 07]

They then decided to face the situation and choose one of the strategies at their disposal: calling on the volcano. It was the *Lin Mal*, the sorceror possessing a hereditary power of speaking to the volcano, who alone could do it. They made an appeal to him, and he said to them:

> Ok. In the next two weeks I will visit them. [Harold Temar emphasizes:] in the two weeks of the visit that Lin Mal had announced, on the 13th December 1913, the volcano came. Fire destroyed everything. [HOS 07]

In November 1913, the Ambrymais were agreed. The community conflicts that were – from their point of view – the cause of the settlers' arrival then ended. The conversion and sociocultural change were perceived as too great a threat. Then, they sought, by all means, to kick the intruders out of the island. Local history does not say if the different misfortunes that had struck Fletcher previously – the fires, the illnesses, the storm, the famine – were of magic or criminal origin. But the Ambrymais claimed at least one of them. They did, they say unanimously, "call upon the volcano". They had recourse to the strength of the sacred fire, to magic, to the power of the ancients to put an end to the situation of profound cultural crisis.

Let us resume: in the first instance, the Ambrymais invited the settlers to come to their island in order to resolve internal conflicts. They were first delighted but then progressively disenchanted, seeing their customs and their moral values disappearing with the conversions. They suddenly saw that a catastrophe was under way, which would lead to the end of what structured their world, that is to say their traditions and rules of community life. They called on the "fire of the volcano" and triggered its eruption to deal with the situation. From a local perspective, the eruption of the volcano did not constitute a catastrophe but, quite the contrary, an ancestral means of defense, a magic wand for managing catastrophes.

7.3.3. *The assymetry of the interpretations*

The year 1913 at Ambrym was not experienced in the same manner depending on who the protagonists were. Depending on whether you look at it from the point of view of the settlers or the Ambrymais, the accounts differ. For the settlers, what made the catastrophe was the (volcanic) *eruption*, whereas for the natives it was the *irruption* (of civilization in their traditions).

In this case study, the interpretative frameworks diverge so far as to be inverted, depending on what structures human milieux, and the regimes of historicity that are engaged. Brutal and sudden (of an eruptive type) or slow and progressive (the type of cultural crisis that the Ambrymais confronted), the hermeneutics of the transformations present a broad spectrum. Depending on the temporality mobilized, the eruption can appear to be a destructive or saving phenomenon, catastrophic or epic, natural or technical. The regimes of temporality in which the actors find themselves mean that a single phenomenon can (1) be integrated in the dynamic of a strategy or (2) emerge after a phase of amazement as something radically uncontrolled, disastrous and undesirable.

At Ambrym in 1913, the protagonists had recourse to two narrative frameworks. Some of the accounts – those of the settlers – hinged upon the volcanic activity, and others – those of the Ambrymais – were linked to the sociocultural impacts of the Presbyterian missions. The eruption of the volcano is concomitant with the irruption of strangers on the island. There was not one event, but two. They simultaneously relate to a syndrome, that is to say a moment where – *a minima* – two frameworks for structuring the experience came to harm, where several milieux are disturbed and lead the collective that inhabit them to react in their specific way.

The population present in Ambrym in 1913 can be divided into two distinct categories of interpretations. The point of view of the settlers expresses a catastrophic volcanic eruption that put an end to a civilizing mission. The point of view of the natives tells the story of a desired eruption that put an end to a colonization perceived as profoundly destructuring. Table 7.1 summarizes the situation.

Event / Population	Volcanic eruption (December 1913)	Colonization (1881–1913)
Settlers	Catastrophe	Civilizing strategy
Ambrymais	Expulsion strategy	Catastrophe

Table 7.1. *Ambrym, 1913: assymetry of the predicative regimes*

To summarize, the population concerned is composed of settlers and Ambrymais. The event links a volcanic eruption and a colonial process. What stood out as a volcanic catastrophe for the settlers corresponded to a traditional mode of resolving social crises for the Ambrymais. What was understood as a saving, civilizing enterprise by the missionnaries was resented as a cultural catastrophe by the natives. There is an asymmetry in the interpretations.

7.3.4. *What makes a catastrophe?*

Depending on whether one places oneself at the standpoint of the "established" inhabitants or the "strangers", what makes a catastrophe for one is a strategy for others. There is an asymmetry of perspectives or, more precisely, asymmetry in the predicative regimes. At the first stage of analysis – basing it on the sole case of Ambrym – we can put forward the hypothesis that a catastrophe is something that derails the plans of individuals and societies, that which does not enter into a strategic framework: for the Ambrymais, the eruption was wanted; for the settlers, it was endured. For the settlers, colonization was considered as civilizing and saving, volontary and positive; for the Ambrymais, it was endured and unwanted, lived as destructive and generated a fear of collapse. Each of the two parties reacted depending on the registers of their own sensibilities and on the registers of their actions, whether symbolic or material. Each felt a fear relative to the non-voluntary character of what confronted them. Each evaluated the danger of the hazard in the light of their field of respective eco-techno-symbolical competence.

The settlers feared the eruption because they considered it to be a self-governing phenomenon, independent of their will. The Ambrymais were gripped with profound anguish at the idea that their traditions and ancestral customs might be changed because they considered them immutable. In the two cases, a danger was experienced when it touched on what the individuals thought of as independent of their will, whether that stands out as (1) the laws of nature (e.g. the tectonic of plates) or (2) social rules (e.g. the alienation of ancestral customs). For each, the ways of being in the world correspond to particular domains of vulnerability, and to privileged registers of pain. The latter also define, by contrast, dimensions of resilience. We thus observe at Ambrym in 1913 a propensity to live effectively the cultural changes on the part of the settlers by integrating them into strategies in which they could act. For "white men", this implies a horizon that can be qualified as "modern", where social changes are perceived as positive, indeed necessary. On the part of the locals, the idea of a human influence on the seismic environment – notably the existence of magic strategies called up through rituals – makes effective the possibility to live with a minimum of suffering on an island with strong volcanic activity. The danger that an individual faces is evaluated in the light of a field of competence [DOU 82, DUC 82], this – material or symbolic, rational or magic – is defined by a certain type of capacity for control imputed with regard to the environment.

In summary, hazards are perceived as catastrophes when they are not integrated into a strategy. Every type of strategy is necessarily relative to a field of competence. The "white man" (*waet man*) fears large-scale environmental phenomena because they believe that controlling these is outside their capacity of action, just as the Ambrymais fear changes that affect their customs since these have, from their point of view, an ancestral legitimacy that exceeds their capacity for judgment. The milieu of the settlers is structured on a essentialist mode of action (a material efficacity), whereas that of the Ambrymais rests upon a magical mode of action (a "symbolic efficacity" in Lévi-Strauss's sense [LEV 62]).

Between action and judgment, the domains of intervention are not the same from one cultural milieu to another. In the first case, it is assumed that it is not possible to modify certain natural phenomena (earthquakes); in the second, it is assumed that one should not change cultural laws (traditions). In the first case, the constraint is considered as being of a material nature; in the second case, it is regarded as being of a moral and ethical nature. In the first human milieu (that of the settlers), the range of human initiatives on what surrounded them arose from the secular world; in the second case, it was ritual (magic). Thus, in the two types of milieux, what affects something that cannot be acted upon is considered as catastrophic. In effect, for each of the protagonists, what makes the catastrophe affects what they consider as "natural", understood here as what is not modifiable under the effect of their own actions (whether technical or symbolic). The catastrophes would therefore always be "natural" on the condition that this term is not considered in contrast to "cultural", but in contrast to the artificial. From this point of view, to say that a catastrophe is "natural" amounts to think that it is not the result of the action and the intention of those who undergo it. Therefore, there is an asymmetry in the interpretations as to what is a product of nature and what is a product of artifice. Table 7.2 summarizes this analysis.

Event / Population	Eruption (volcanic)	Tradition (Ambrymais)
For the "white man"	"natural" (one *cannot* modify it)	"artificial" ("one *should* civilize savages")
For the Ambrymais	"artificial" (one *can* cause it using magical techniques)	"natural" (one should not modify it)

Table 7.2. *Ambrym-1913 (synthesis)*

Hazards are perceived negatively when they touch upon what makes up the world, on what is culturally assumed as a principle of order and stability. According to the milieu to which these customs belong (Ambrymais), permanence is social, relative to ancestral customs (*nomos*), whereas from the point of view of a "white man's" milieu, continuity is assured by the nature (*phusis*) of immutable laws. Each of the cultural schemes induces a certain type of sensitivity to change. The susceptibility accrued by the Ambrymais in the face of cultural variations, the greater worry of the "white men" in the face of environmental instabilities, can from then on be understood as flowing directly from the manner in which they elaborate their own milieux.

To summarize, at Ambrym, in 1913, the "white men" (*waet man*) came to alter the ancestral traditions of native populations (this arose from their sense of moral duty), but they felt impotent in the face of a volcanic hazard that exceeded their capacity to act (they could do nothing). The Ambrymais took their customs for granted (they should not be modified), but they assume a role in the behavior of the volcano (they believe that they are able to trigger its eruption, notably through magical practices). The hazards are therefore predicted depending on the schemes of actions adopted by the human milieux under consideration. Whatever the milieux, they appear as systemically "natural" (in the sense that they are not modifiable).

Beyond this case study, which presents two radically different populations from a cultural point of view, such asymetries in perception can be encountered within a single society. Religious persons, who understand disasters in their lives by thinking of them as divine signs (e.g. as punishments), create a model for resilience (and correlatively a model for vulnerability) asymmetrical to that of someone who makes scientific models the basis of his/her reasoning. It is important to note that the so-called "cultural" variability does not indicate differences in culture, but ways of making sense of the world which, although they are influenced by social environments, depend ultimately on individuals. The following case shows this variation.

7.4. Edo, 1855

In the second year of the Ansei era, the second day of the 10th moon, at the hour of the boar – being, in the Gregorian calendar, the 11 November 1855 around 22:00 – the city of Edo (the former name of Tokyo) was shaken by a strong earthquake. Its intensity was not exceptional – 6.9 on the Richter scale – but the epicenter, not very deep and situated just under the Japanese capital, caused considerable damage. More than 10,000 inhabitants of the imperial capital, the seat of power for the Tokugawa, were killed. Around 20,000 homes were reduced to heaps of rubble and ashes. Moreover, numerous copies – of the archives relating more than 50 accounts of it – would contribute to durably maintaining the post-seismic state of tension and worry. The damage was not uniform: certain quarters were untouched, some were completely devastated, certain types of buildings resisted the quake and others systematically collapsed. The distribution of damage influenced – and this is a very normal attitude in the face of disasters –

the interpretation of causes and the distribution of instances or people judged responsible. For the case that will concern us presently, the Edoites would notice two singularities.

7.4.1. *Destruction of governmental infrastructures and granaries*

Most of the populated living quarters (with the exception of Fukagawa and Yoshiwara) as well as some residences of the Japanese nobility (*daimyô*) were spared, but the administrative and military infrastructures, linked to power (that of the *bakufu*), were for the most part severely damaged. The contrast was so striking that the Japanese felt that the government was "targeted". This contributed to increasing a social instability swift to overinterpret anything that might be susceptible to consolidating the well-foundedness of the critics with regard to the government. In the first instance, the latter had to face a flood of protestations.

Figure 7.1. *The government, symbolized by the horse without a rider, is brought down by a Namazu (source: [SMI 08, p. 6])*

The first wave of prints criticized the ruling powers with such virulence that the *bakufu* strove very quickly, after only a few days, to forbid their production and diffusion. But the sanctions still had no effect. The anger increased. It was necessary to react rapidly with concrete and visible measures. From the day after the earthquake, the municipal council (*machigaisho*) of Edo specified a series of measures that ought to be put in place to help gain control of the situation.

Figure 7.2. *Makeshift government shelter (source: [SMI 08, p. 2])*

The restrictions were numerous and the instructions were divided into nine major areas [SMI 08, pp. 3–6] :

– to distribute food to the living, for example balls of rice, in order to relieve the famine situation in which part of the population found themselves;

– to put in place makeshift shelters to house the homeless;

– to send medical help urgently to take care of the injured;

– to convene the main distributors of merchandise with a view to building up essential stocks;

– to convene those responsible for transport companies to bring competent workers from other regions;

– to forbid buying and selling on the black market;

– to control the increase in the cost of goods and salaries;

– to install patrols of police officers (*yoriki* and *dôshin*) to assist the population and supervise the measures for alleviating the disaster;

– to consolidate the legitimacy of the district officials (*chô*).

This ensemble of measures would function very well which, after two months of considerable investments, enabled the government to calm their critics [SMI 08, p. 10]. This point is important – i.e. the financial engagement of the Japanese government – which would be added to the framework of the general interpretation of the catastrophe: the circulation of the "vital breath" (*ki*) of which the riches formed a part.

Among the other damages that appeared remarkable, the Edoites noted the almost systematic collapse of fortified granaries. With their thick walls constructed to prevent theft, these buildings, because of their rigidity, are very sensitive to earth tremors. Despite the supports, which increased their resistance, many granaries collapsed. There too, this destruction was not perceived as being a work of fate, but as pointing the finger at a social class which, since the mid-19th Century, had been the object of an intense envy and growing jealousy: the merchants. The prints, that is to say the media of the era, made them a laughing stock, showing them in the process of defecating their gold. The captions that accompany the images give a cosmological explanation for this scourge, which affected all the commercial class.

From the point of view of traditional Japanese society, the merchants fulfilled their function from the moment when they redistributed their richess and enabled them to circulate, especially the gold coins (*ryô*). According to Japanese cosmology, gold is effectively considered as "an essential element that should circulate". By accumulating it, the merchants were not only capitalizing their assets but also causing the "vital breath" (*ki*) to stagnate.

Figure 7.3. *"The treasures of the vessel, The Miser emptied" [mochimaru takara no defune]. In the caption: "The money goes into the pockets of the workers and is distributed throughout society, which is heading toward a restablishment of the economic health of the nation" (translation: [LUD 06, p. 1059]), source: [BUT 99, p. 30].*

Figure 7.4. *"The disease of the millionnaires' metal" (chôsha, kane no yamai). A Namazu, resting on the back of merchant, makes him vomit his gold. On the right-hand side, a fortified granary can be seen, having fallen victim of the earthquake. On the left, two merchants defecate gold (source: [LUD 06]).*

From this perspective [SMI 06], all capitalization triggers something like "constipation" of the social body, and the earth tremor appeared to be "a powerful laxative of therapeutic virtue". The events of the 11th November 1855 were perceived as something that triggered the forced restablishment of the circulation of riches, whether they were accumulated by the merchant classes or by the government.

7.4.2. *Incooporating the rise of globalization*

The government was moreover criticized over a question that was hitherto unprecedented for the archipelago. At the end of the 19th Century, a new geopolitical order came to light, commercial and cultural exchanges were structured on a global scale. The industrial and political revolution accelerated the flow of people, ideas, techniques and objects, and trade systems became widespread, over the entire planet. Globalization was underway. Indeed, for nearly three centuries, the archipelago had been "in quarantine". It had closed its frontiers to all forms of migration – no one went in, no one went out – even though, for several decades, foreign boats regularly came to its bays to solicit trade. More demanding and menacing each time, Russian, Dutch, French and, above all, American governments put pressure on the *bakufu*. They requested, then demanded the opening of the territory. Commodore Perry's black ships, at anchor in the bay of Tokyo, were the most famous and, and with the firing power of their Paixhans canons, the most feared. There too, but on a geopolitical scale this time, it was possible to interpret this isolationist politics (*sakoku*) as a stagnation of the "vital breath" (*ki*). In other words, governmental politics with regard to globalization could also, from a local point of view, have triggered the volcanic eruption. Analysis of the media treatment of the event points toward this direction/meaning.

With a few very rare exceptions, all the prints published during the month that followed the eruption of 1855 made no use of dragons – until then traditional figures associated with earthquakes – but to mythical catfish, the *Namazus*. One series of prints, which depicts the *Namazus* as having a striking resemblance to Commodore Perry's black ships, is significant [NOG 04, SMI 06].

Figure 7.5. *"The giant cat-fish – the prosperity of Edo" (ô-namazu Edo no nigiwai). This representation of the Namazu is reminiscent of Commodore Perry's "black ships" (source: [BUT 99, p. 28])*

The catfish represent a principle of disorder capable of integrating into a single metaphor the turmoil of a mythical beast capable of triggering earth tremors and Commodore Perry's "black ships" (黒船, *kurofune*). In 1855, the change from the depiction of seismic disasters using dragons to their depiction by *Namazus* signifies the arrival of a new category of menace – the globalization of commercial and cultural exchanges. In this regard, it is significant that in 1863, Japan would sign the free trade agreement and that in 1868, this country would begin a voluntary "enlightenment" (*Meiji*) where they would aim to participate fully in this new cultural order, of which Japanese society formed only a part.

7.4.3. *Victims and profit-makers*

The rational based on an economy of *ki* (vital breath) is a cosmological scheme for the general interpretation of disasters in the Japanese milieu. It does not come down to circulation of a pecuniary kind but is depicted by analogy with all exchanges, for example the distribution of good or bad fortune. This ambivalence is also present in the numerous prints. It is clearly illustrated by one of them, the most famous.

A varied population is shown here (courtisans, clients, inhabitants of the quarter and children), who attack at the same moment, each according to his means – some with hairpins, some with fists, one with a cane, one with a club, a knife, a head-rest or a shamisen – a giant catfish (*Namazu*). He warns them: "I am delighted that all these beauties are climbing on my back. If others come again, I could shake again". But the image also shows (Figure 7.6) carpenters and firemen who, having profited from the event, run over and ask for mercy, for its agressors to leave it alive. "They cry: 'Wait! Wait! Wait!', 'Stop! Stop! Stop!' 'Hey, Hey, Hey! Don't hit it like that! Don't hit it like that! Don't hit it like that!'".

Figure 7.6. *"Controlling the giant Namazu of New-Yoshiwara" (Shin Yoshiwara dai namazu yurai), Miyata and Takeda, 14–15, impression no. 46 (source: [CAR 08, p. 157])*

This calls the obvious to mind: catastrophes are not only catastrophic. For some, the disaster constitutes an opportunity, a real windfall. It opens an immense site for reconstruction, which promises work for many years.

7.4.4. *Tears and laughter*

In Edo, in 1855, mediatization occurred in the form of images commentated and diffused through the means of a form of itinerant theater (*kamishibai*, 紙芝居). As today, when newspapers report major events, this press experienced a dazzling success. From the 13th November, the prints were diffused and sold in large numbers all around the devastated town and its surrounding area. After a few weeks, more than 400 different prints were competing on the market. More than 10,000 copies of some exemplars were printed. As today, the diffusion of information was directly linked to a commercial and promotional cost. The prints formed part of a strong economic logic. Some productions were positively advertisements.

There are therefore broad similarities between the media treatment of the catastrophe in historic Edo and its treatment today in the *mass media*. However, an important difference in dramatization, that is in creating a narrative of what has happened, distinguishes them. Contrary to current media, they do not handle the event in a merely tragic tone, nor center on the catastrophic. Very much to the contrary – as all those who have studied the body of prints have pointed out – they do not lack humorous features.

The overall spirit in which the disaster is treated is humorous. The prints draw attention to the strange or freaky aspects of the event. "All depict one or several catfish quite commically", summarize Butel and Griolet [BUT 99]. They do not tend to reinforce the disastrous side to what has happened but, much to the contrary, they emphasize anecdotes that are amusing, and aspects that might provoke laughter. Far from consisting of a series of varied facts, each more depressing than the next, the printed sheets were created in a mood of pleasure and entertainment, more precisely, in a tone of irony and sarcasm, satire and derision.

7.4.5. *Synthesis*

The complexity in the interpretation of this catastrophe has been woven together through the acts of a *Namazu*. This simultaneously represents:

– something perceived as a geological rule: seismic activity;

– the presence of a new form of circulation on a planetary scale: globalization;

– the "poor choices" followed by the Japanese government: refusing to open the territory to foreigners and accumulating riches;

– the tendency of traders to allow their gold to "stagnate" in granaries;

– the varied distribution of the impacts of the catastrophe, a source of misfortune for some and good fortune for others;

– treatment of the disastrous and tragic nature of the event, as well as its humorous and ironic dimensions.

The earthquake is thus interpreted in the light of a logic that links the immediate effects (misfortunes and destructions), the mid-term effects (reconstruction and work) as well as the long-term effects (reconstruction of social dynamics and "vital breath").

This is why the *Namazu* is not only represented as a uniquely harmful and disastrous force, but also as an archaic, primordial energy that obliges Japanese society to respond to a long-term logic: the resultant destructions force the liberation of the flux accumulated by society (the riches of the merchant classes and the government) or the fluxes they refused (globalization), in order to remobilize vitalizing forces (those of laughter and reconstruction) in the face of adversity (the deaths and destructions that had been caused). From this perspective, human existence supposes (and relies upon) a coexistence (*kyôson*, 共) with natural catastrophes [AUG 12]. The seismic shock is perceived as a fundamentally positive force, which "rectifies the world" (*yo-naoshi* 世直し), a real kernel of a principle of profound harmony (*wa*, 和) which – although destructive – "is sometimes accompanied by positive counterparts for the survivors".

Figure 7.7. *"The namazu deliverer of medicines" (Furidashi namazu-gusuri). The texts indicate that the earthquake is a medecine that restablishes the flow and circulation of the money collected in warehouses, restores warmth in injured hearts, takes care of poverty, reduces laziness and prevents the harmful effects of a luxurious life (source: [SMI 06, p. 1061]; text: [LUD 06])*

7.4.6. *Conclusion: the fundamental ambivalence of catastrophes*

The graphic treatment of the volcanic eruption that struck Edo in 1855 constitutes an example of the media treatment of catastrophe lacking in catastrophism or complacency. This case furnishes all the elements that illustrate what a catastrophe can consist of in a concrete sense: although one region may be hit by disaster, only a portion of the life forms that make up this milieu is affected in a harmful manner.

Figure 7.8. *"Fear of Kashima", feast in honor of the Namazu (source: http://shinku.nichibun.ac.jp/namazu/sakuhin.php?id=76)*

In fact, what happened on 11th November 1855 in Edo was not experienced unilaterally as a tragic event. The prints tend to represent it in a complex manner: as a resource (for the artisans, monks, firemen, etc.), entertainment (for the prints' readers), impediment (for the government and the victims) and risk (seismic and political). This media treatment depicts the impact of this disaster in a manner at once serious and light, taking an interest in the victims as much as in those that profited from the *tabula rasa,* that is to say, those for whom the destruction was an opportunity. They give the impression that what has happened retains a fundamentally ambivalent character, an ambivalence even at the heart of a principle of "coexistence" (*kyôson*).

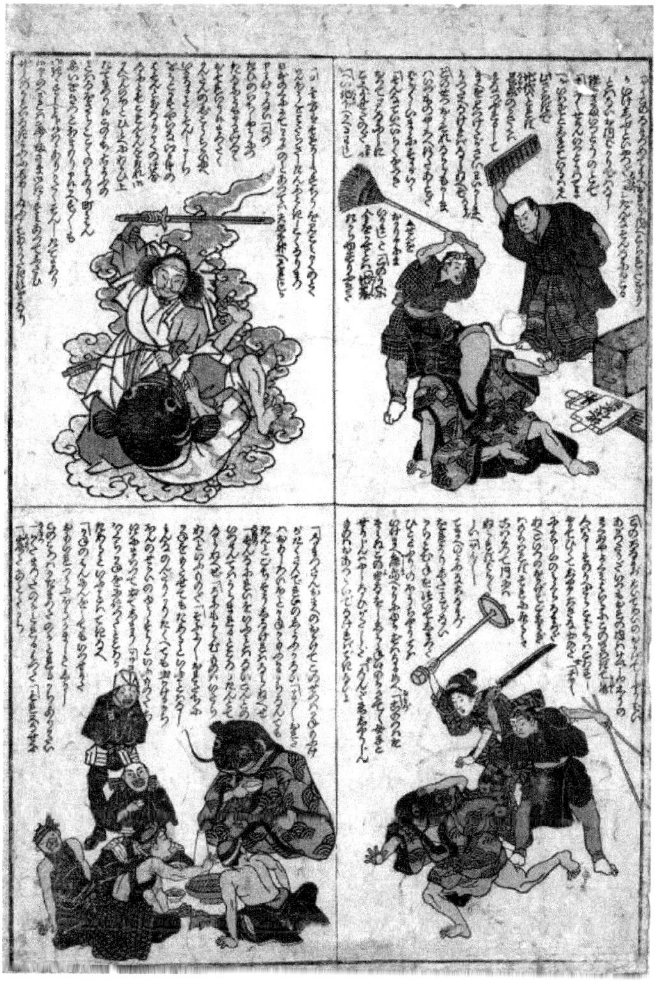

Figure 7.9. *"Short stories of the earthquake" (地震小咄) united on the same plate: the protectors and victims of the Namazu (source: 鯰絵コレクション http://shinku.nichibun.ac.jp/namazu/sakuhin.php?id=19)*

In a fairly objective manner – and this despite the lack of discernment – the creators of the prints were able to place in the images both the benefits that were created as much as the impediments that had to be overcome, the situations that lent themselves to comedy as much as the tragic scenes. This way of representing the event incites and engages the readers to look past the evidence of the disaster and the stupor that it generates: although they are not denied, the desolation of the city in ruins and the remains of victims charred or crushed under the rubble are not at the heart of the commentary nor at the center of the representations. In fact, a catastrophe has no center, but rather an epicenter, it branches and spreads in a varied and complex manner, and it does not give rise to a simple consensus, but to a controversy. In fact, the concrete impact of what happened at Edo in 1855 has followed extremely diverse trajectories, and not all have been disastrous. This branching of the impact is not unique to this catastrophe; all major disasters have a "horizontality" (in space and depending on the protagonists) and a "verticality" (in time, from the event to long-term societal and geological changes) distributed in a non-isotropic manner.

To leave space for this ambivalence, to learn to depict it in the face of everything that is presented as catastrophic (and not as a catastrophe), it is sensible, as the prints that we have just studied have been able to do, to systematize the depiction of "happenings" according to the various protagonists (victims and profitors) and the levels of the impacts (on the scale of the event and the historical and cosmological movements under way). This operation, to which we are now going to return, was initially formalized by Augustin Berque by means of what he called "medial categories":

> The approaches of the ecumene can be classed in four categories: resources, impediments, risks and pleasures. In general, however, all reality participates in variable proportions in each of these four realities. For example, on a mountain, the snow is at the same time a resource for a promoter of ski resorts, a constraint for the farmer, an avalanche risk for everyone, and pleasure for the skier. [BER 11]

This is one of the hypotheses of Berquian mesology: everything that happens is "grasped" as (something) and this "grasping" can be organized into four medial categories: (1) risk, (2) impediment, (3) pleasure and (4) resource [BER 00, pp. 149–150]. These categories define what Augustin Berque calls the "medial holds". When joined together, they form the "mediance" of an individual: the way in which a human being gives meaning to what happens to him. In this meaning, for Berque, there is an emergence of meaning in the measure where an individual views his surroundings at the same time as a risk, a pleasure, an impediment and a resource. These categories together form a "geogram" [BER 00], which describes the ensemble of the understandings assumed with regard to a thing or an event. In other words, by working the depiction of a "happening" according to basic "ecumenous motives", we produce its "geogram".

This brings us back to take into consideration all the values (pessimal[1] and optimal) linked to the material impact of the disaster (destruction and production) just as much as its immaterial impact (symbols and subjets). It is a question of placing two couplings in relation to one another: material–immaterial and optimal–pessimal. Table 7.3 provides a simple overview of them [MOR 13].

Disaster	Optimal	Pessimal
Material	Resource	Constraint
Immaterial	Pleasure	Risk

Table 7.3. *Geogram of a disaster as a clutch of structural material–immaterial coupling and systemic optimal–pessimal coupling*

In the above table, the elements are read as follows:

– a resource is something that enables the physical and organic (that is to say material) milieu of an individual to be maintained and promoted;

– an impediment is something that has an effect and is physically present (it is material) which impedes the individual's relationship to what surrounds him/her in such a way that he/she cannot engage with the latter in an optimal manner;

1 The term "pessimal" is borrowed from J. Von Uexküll [UEX 56], and means the opposite of optimal. Thus, the "optimal conditions" mean a milieu favorable to development (of a life form, of society, of existence, etc.). On the contrary, "pessimal conditions" mean a milieu harmful to development (of a life form, of society, of existence, etc.).

– a pleasure is something that enables an individual to enjoy (morally and symbolically) the milieu he/she inhabits;

– a risk is immaterial (since it applies to disasters that are certainly material but that have not yet occured, which do not therefore have a concrete existence), and leads to a potentially "destructuring" relationship (in the sense of Watsuji, that is to say destructuring for its mediance).

7.4.7. *An analytical tool: the ideogram of a catastrophe*

The anthropological approach to disasters requires a means of illuminating the concrete reality of a catastrophe, that is to say its meaning in relation to the milieu where it occurs. Something that is experienced as catastrophic equates something in which the meaning is lacking because this quadruple unpacking has not been – or has only partially been – operated: the victims can barely envisage the positives in the event, its motivating, structuring, beneficial and serendipitous dimensions. From this perspective, the catastrophic is something that remains a state of menace, fear and anguish, when the protagonists' relationship to it remains stuck solely at the level of risk and impediment. Indeed, any reading of disasters that pays attention only to risks and impediments will end in catastrophism and will lack "symbols" with which its overall meaning can be grasped. Speaking of "symbols" enables us – by analogy with the Japanese or Chinese ideograms – to put forward a graphic version of catastrophes. An ideogram is in effect composed of "symbols", which, when combined spatially, suggest ideas. For example, the "symbol" for water associated with that of fire gives – in Japanese – the idea of a catastrophe. Each symbol conveys a specific meaning, a trajectory for interpretation of what has happened, but it is together that they form a general idea, the overall character of the catastrophe.

It is possible to develop a graphic form for the understanding of catastrophes. It enables "connections to be made visible" [WIT 92] between the disaster and its concrete reality (which is at once ambivalent and diachronic). Such an "ideogram" of catastrophe can be operated by linking the four "symbols" that constitute the four medial categories. The first represents the risks incured, the second represnts the resources that are produced, the third reprents the pleasures that can be taken from it and the fourth represents the impediments created. When a catastrophe is envisaged according to the four "symbols", an ideogram is constructed: its meaning relative to a given human milieu (that is to say its mesological meaning).

	As a RESOURCE	
As a RISK	CATASTROPHE	As a PLEASURE
	As an IMPEDIMENT	

Table 7.4. *Ideogram of a catastrophe depicted according to the medial categories (risk, pleasure, impediment, resource)*

7.4.8. *Example*

By applying this analytical tool, it is possible to suggest a reading of the case study analyzed above. We thus obtain, in a synthetic form, the following ideogram:

		RESOURCE		
		For artisans and reconstruction		
RISK	Earthquake and globalization	Eruption at Edo (1855)	Unusual and humorous stories	PLESAURE
		For the victims		
		IMPEDIMENT		

Table 7.5. *An ideogram of the volcanic eruption in Edo in 1855 (written version)*

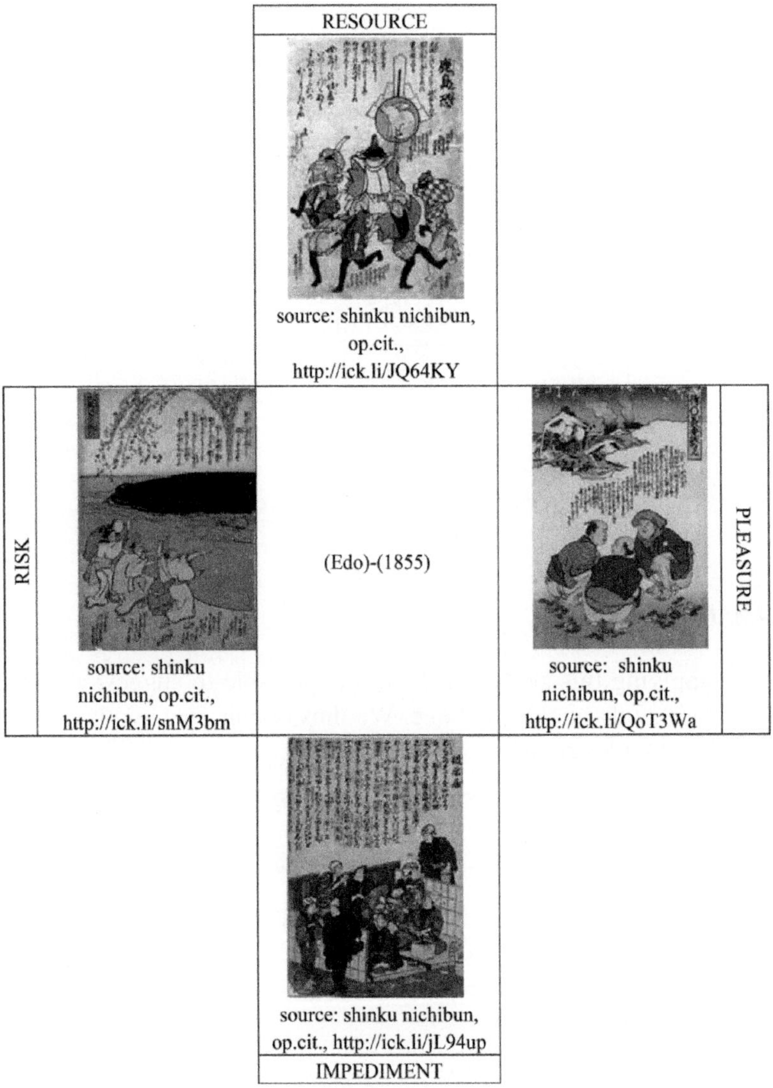

Figure 7.10. *A geogram of the volcanic eruption in Edo in 1855 (pictural version)*

These ideograms synoptically present the conclusions of an analysis of the 1855 Edo earthquake's impact on the Japanese milieu. What happened that day was not treated by the media of the era in a unilaterally catastrophic manner. It was also considered as a windfall (a resource), as the hinge of a plot subject to the enjoyment of narration and pictorial depiction of a "humorous" character, and as a factor that forced the authorities and the merchants to relinquish their fortunes to meet the essential needs of the impoverished survivors. From a mesological perspective (that is to say, one that takes account of the human milieu concerned), this interpretation can be translated as an "ideogram of catastrophe". This can constitute a tool that displays, in a synoptic manner, the ambivalence that appertains to all major disasters: its concrete and complex expression. It also allows us not to give way to the attractions of the ever-present catastrophism and to observe carefully the resources and benefits, attractions and desires, pleasures and hunmorous aspects which, whatever one says, irrevocably remain the result of disasters.

Figure 7.11. *Print by Katsuhiko Hokusai called "the Waves"*

7.5. Tsunami or tidal wave?

Let us observe in detail this famous print (Figure 7.11) that the artist Katsuhiko Hokusai painted in 1830 and 1831. Extremely fine droplets form fractal details in the form of claws, which together make up an immense wave. This threatens to engulf two canoes in which 20 tiny Japanese fishermen lie, still gripping their oars. An effect of perspective adds a cosmic dimension to these boats in danger of being wrecked: Mount Fuji can be distinguished in the distance; an ancestral symbol of the permanence of Japanese culture, it appears, too, to be in the process of being dragged into the heart of the waves. In this famous print by the painter Hokusai, the Japanese world is considered as an ensemble where whatever disturbs a microcosm (the individuals in their boats) also threatens the macrocosm (the geographic). This principle of correspondance between the micro and the macro (of which we have shown evidence in the previous case) is represented here on the brink of sinking "under an enormous wave offshore from Kanagawa" (litteral translation from the original title). The print represents a situation of impermanence (in Japanese *mu*), where the life of man compared to Mount Fuji appears as similar to that of the petals of a cherry in the face of the spring wind or as the finest droplets of this immense wave in the face of the swell.

The formal beauty of this work alone could have sufficed to ensure its diffusion at a museographical and academic level. But it is as an emblematic symbol of tsunamis that it has only recently, in the mass media and on digital networks, undergone its most significant diffusion. Today, it is one of the most well-known prints in the world, and when we speak of a tsunami, it is usually this image that comes to mind. Yet, "Under the wave of Kanagawa" does not show a tsunami. A shift in interpretation has been introduced between what Hokusai has shown (a "large wave", in the Bay of Tokyo, offshore from the city of Kanagawa) and the mediatic use that we know of this image today (as an iconic symbol of tsunamis).

This shift reveals a major cultural difference regarding the relationship of a tidal wave with natural disasters in general. To really understand this difference, it is sensible to compare the two words presumed to mean a single phenomenon: "tidal wave" and *tsunami*. This homology is not self-evident. We will explain. First, in what way does a tidal wave resemble a tsunami and, second, in what way is a tsunami not like a tidal wave?

From the geological point of view, there is no difference between the realities designated by the terms "tidal wave" and "tsunami". Both result from a telluric phenomenon (e.g. earthquake and volcanic eruption) or something similar (e.g. landslide and meteorite shower), leading to a sudden displacement of a significant mass of water (lake, seas and oceans). The waves triggered by the non-telluric phenomena (wind or phenomena of resonance and corresponding wave phases) have their own terminology: "storm" or "rogue" waves. The distinction does not arise from an objective difference; it does not arise from the phenomenon "itself", that is to say, considered as uniquely of a mechanical (tectonic) order. The distinction is of a mesological order, that is to say, it relates to the impact of this phenomenon on the human milieu. A tidal wave can have no human consequences, but not a tsunami. A tsunami is a tidal wave that affects a community of people. A tidal wave that occurs in an uninhabited zone (e.g. off Greenland or in the open sea) is not a tsunami, whatever its size. In this case, from the Japanese point of view, it is simply a question of a "large wave". Moreover, even in cases where people are present over a generally uninhabited zone (e.g. a group of scientists or hunters on an expedition to Greenland, or – as Hokusai represented – sailors at sea), it is not a tsunami. A tsunami is a tidal wave that penetrates an ecumene, that is to say a zone inhabited by mankind. It does not only threaten their existence, but a milieu in its entirety. Because of this, since it relates to a variety of protagonists and occurs in a social temporality, it creates a divergence in interpretation that creates an argument and controversy on a social scale.

The word "tsunami" is written in Japanese by means of two ideograms, each combining two symbols. The first ideogram means "large wave" and the second ideogram means "port". Etymologically ("etymographically" we should say), a tsunami is therefore "a large wave that flows inside a port", that is to say a tidal wave, a storm wave or a rogue wave that penetrates an inhabited zone and is experienced by a human community. This definition is neither objective (describing a purely mechanical phenomenon) nor subjective (due to the interpretation of a single individual); it is "trajective" (that is to say alternance of a process of objectivation and subjectivation). It is dependent on the physical (a large wave, whatever its cause) and social (across differences in interpretation of an ensemble of protagonists) aspects of the event.

In other words, the word "tsunami" conveys a general scheme of apprehension and interpretation of major disasters: as a "meeting" between nature and societies or, more precisely, as an essential motif for the coexistence of the environment and human milieux. Let us look again at Hokusai's print. Does it represent a tidal wave or a tsunami?

– It may have been triggered by a telluric phenomenon, in which case it depicts a tidal wave;

– it could also originate from a resonance phenomenon in sea waves, in which case it represents a "rogue wave";

– it may have been shaped by the wind, in which case it is a storm wave;

– on the other hand, it is not entering an inhabited zone; therefore, in any case, it is not a tsunami.

We have an additional sign, inscribed from top to bottom in the rectangle, Figure 7.11: "Enormous wave offshore from Kanagawa". It could feature an exceptional wave or a roller that makes up one of the "spots" sought out for pleasure by surfers. We do not know if Hokusai is depicting the daily life of Japanese fishermen and seeking to show their intrinsic courage, or if he is illustrating a rare event they have to confront on this occasion. Hokusai leaves space for ambivalence, that is to say for the quality that – at bottom – significantly characterizes major disasters. It is also here that the essential difference lies between the meaning of the words "tidal wave" and "tsunami". The first is a normative concept based on factors independent of human milieux (the physical intensity of the tectonic phenomenon, the height and speed of the wave). The second integrates a social percept, which takes into consideration the impact of the phenomenon on a region inhabited by a human group (a milieu, an ecumene, etc.). Given that, the ways of living in the world vary from one society to another; the use of the word "tsunami" – when one wishes to use it in other languages – will therefore vary as well. But, in all cases, it cannot be reduced to objective conditions; it cannot be conflated with the term "tidal wave". Or, indeed, this would be to deny this other way of evaluating the social vulnerability which does not consist of "natural" phenomena, on the one hand, and "cultural" phenomena, on the other hand, but a "mesological" modality where nature and culture form an inseparable couple which, like all couples, seeks to coexist (*kyôson*) together (and not separately).

7.6. Conclusion

Meaning is never clear and never attained, particularly where taking account of catastrophes is concerned. It is only possible to render them "sane", that is to say acceptable to the degree that one can continue to exist and to wish to live. The main challenge of the human milieux is twofold: physical and mental self-preservation. But the cosmologies vary: some consider bodily wounds to be more fundamental than moral outrages; others believe that abstract problems are more dangerous than material disorder. In all cases, the challenge does not arise uniquely from a desire to know the general (objective truth) or the one that an individual constructs (subjective truth); the first challenge is to be able to live with the major disasters with which the history of our milieu is dotted. The Japanese have a culture that enables them to inhabit a highly seismic zone without this affecting the foundations of their social consensus. The Amazonians have a mode of life that accomodates annual flooding that covers five times the built-up area of France, and this is not catastrophic. Each society moulds the world in such a way that it appears as stable and ordered, that is to say in such a way that the disasters that affect it appear as sane.

The evaluation of vulnerabilities to and the impacts of a disaster is a complex task in which it is sensible most of the time to temper any form of catastrophism. This latter consists of mixing the benefits and the risks, the potential and the vulnerabilities. Indeed, the anthropology of catastrophes is not the anthropology of the catastrophic. Pondering what makes up an event (an earthquake, a tsunami, a flood, etc.) and observing the ensemble of its consequences, harmful as well as benefical, is not the same as tending to be systematically pessimistic and normative. This type of attitude tends to displace attention toward a certain category of people (the victims) and the associated damage (the material costs). Yet, it is a question, in order to understand the overall spectrum of the consequences of a disaster on the scale of a human group (a community, a society, a nation, an ecosystem, etc.), of considering the ensemble of its incidences. In this sense, it would be convenient to depict this complexity systematically and to envisage the disasters not merely in terms of risks and impediments, but also as resources (e.g. the benefits induced by a *tabula rasa* favorable to land building work, and to generating jobs in the field of reconstruction) and pleasure (e.g. the comical situations that have been experienced, as well as the spectacles, films, stories, etc., for which they are the inspiration).

An anthropological approach to catastrophes urges us to consider vulnerabilities and major hazards as not arising from a classical logic – even dialogical and "complex" [MOR 73], integrating the "included middle" [BER 00], a rythmicity in "mixing" [LAP 01], or a "multiplicity" [DEL 80] – which can be carried out by connection *logos*, but from a graphic logic which examines and compares the disparities of the impacts, the divergences in the interpretations, the differences in the milieux and deviations between temporality regimes. In this sense, it is sensible to link visual schema (which works on the relationship between the signifier and the signified) to the discursive treatment of catastrophes (which is built upon an arbitrary sign, that is to say the relationship between the signifier and the signified). In other words, there is not really an anthropological "point of view", but a succession of presentations of "effects of perspective" which lend themselves to a grasp of the ways in which particular points of view are formed and developed. The comparison of these "effects" involves working ahead of discursive logic alone, at the level of what Lyotard [LYO 88] called the "figural".

An ideographic presentation of catastrophes enables them to be treated in a figural manner, that is to say in such a way that the meaning does not follow a linear logic, but spreads and contracts, comes and goes (from one graphic region to another), assembles and rises (from aspects of the figures), sequences and contemplates (from more or less prolonged moments of attention), and joins and separates (meanings). In other words, the figural space, because it gives priority to looking over listening, tends to favor the activity of the receiving subject over that of the emitting subject. In doing this, it displaces the action of interpretation. It is no longer the analysis (the author's interpretation) that takes priority, but the "catalysis" (the mobilization of the interpretative capacity). Indeed, the skill of overcoming divergences in perception and reception – of reorganizing the social consensus and the relationship with the environment in times of controversy – is at the heart of relationships to major disasters, whatever the societies concerned. Thus, in the face of catastrophes, it is not the trajectories of the interpretation of disaster that prevail, but the interpretative systems of action to which they belong. It is not their explanation that takes priority, but the ability to have a "grip" on them: to be able to take hold of them (a technical register), understand them (a symbolical register) and also to have an understanding of the other forms of grasping them (cultural register).

7.7. Bibliography

[AUG 11] AUGENDRE M., "Risques et catastrophes volcaniques au Japon: enseignements pour la géographie des risques", in NOVEMBER V., *et al.*, (eds.), *Habiter les territoires à risques*, PPUR, Lausanne, pp. 185–206, 2011.

[AUG 12] AUGENDRE M., *Kyôson, 共存: la coexistence*, Nichibunken, Kyoto, 2012.

[BEC 01] BECK U., *La société du risque. Sur la voie d'une autre modernité*, Aubier, Paris, 2001.

[BER 90] BERQUE A., *Médiance: de milieu en paysage*, GIP Reclus, Montpellier, 1990.

[BER 00] BERQUE A., *Ecoumène. Introduction aux milieux humains*, Belin, Paris, 2000.

[BER 11] BERQUE A., Prises et déprises (A. Berque)/catastrophes (Y. Moreau), 28 March 2011. (Consulted 23rd November 2012 on Mésologiques.fr: http://ecoumene.blogspot.fr/2011/01/prises-et-de-prises-berque-catastrophes.html).

[BON 13] BONNEUIL C., FRESSOZ J.B., *L'événement Anthropocène. La Terre, l'histoire et nous*, Le Seuil, Paris, 2013.

[BOU 11] BOULEAU N., *Risk and Meaning*, Springer, New York, 2011.

[BUT 99] BUTEL J.M., GRIOLET P., "Histoires de poissons-chats – les images du grand séisme de 1855 à Edo", *Ebisu*, vol. 21, pp. 17–34, 1999.

[CAR 08] CARTWRIGHT J., NAKAMURA H., "Tsunami: a history of the term and of scientific understanding of the phenomenon in Japanese and western culture", *Notes & Records of the Royal Society*, vol. 62, pp. 151–166, 2008.

[DEL 80] DELEUZE G., GUATTARI F., *Mille Plateaux. Capitalisme et Schizophrénie*, Minuit, Paris, 1980.

[DES 05] DESCOLA P., *Par-delà Nature et Culture*, Gallimard, Paris, 2005.

[DOU 82] DOUGLAS M., WILDAVSKY A., *Risk and Culture*, University of California Press, London, 1982.

[DUC 82] DUCLOS D., "La construction sociale du risque: le cas des ouvriers de la chimie face aux industriels", *Revue Française de Sociologie*, vol. 28, no. 1, pp. 17–42, 1982.

[FLE 89] FLETCHER R.J., *Lettres des mers du Sud*, Minerve, Paris, 1989.

[GID 94] GIDDENS A., *Les conséquences de la modernité*, L'Harmattan, Paris, 1994.

[HOS 07] HOSNI S., Les voix des volcans: enquête sur les différents modes de faire la connaissance du volcan d'Ambrym, PhD, LAS, EHESS/ Collège de France, Paris, 2007.

[LAG 81] LAGADEC P., *La civilisation du risque: catastrophes technologiques et responsabilité sociale*, Le Seuil, Paris, 1981.

[LAM 05] LAMB R., *Saints and Savages*, William Blackwood & Sons, Edinburgh and London, 1905.

[LAP 01] LAPLANTINE F., NOUSS A., *Métissages, de Arcimboldo à Zombi*, Pauvert, Montréal, 2001.

[LER 64] LEROI-GOURHAN A., *Le geste et la parole*, vol. 2, Albin Michel, Paris, 1964.

[LEV 62] LÉVI-STRAUSS C., *La pensée sauvage*, Plon, Paris, 1962.

[LUD 06] LUDWIN T., SMITS G., "Evolution of the catfish (namazu) as an earthquake symbol in Japan", *Seismological Society of America*, 2006.

[LYO 88] LYOTARD J.F., *L'Inhumain. Causeries sur le temps*, Galilée, Paris, 1988.

[MAR 14] MARSHALL P., "The recent volcanic eruptions on Ambrym island", *Transactions and Proceedings of the Royal Society of New Zealand 1868-1961*, vol. 5, pp. 387–391, 1914.

[MIL 89] MILLER G.J., "A history of church planting in the Republic of Vanuatu", *Live*, Presbyterian Church of Vanuatu, Port Vila, vol. 6, 1989.

[MOR 73] MORIN E., *Le paradigme perdu: la nature humaine*, Seuil, Paris, 1973.

[MOR 99] MORIN E., *Les sept savoirs nécessaires à l'éducation du futur*, UNESCO, Paris, 1999.

[MOR 08] MOREAU Y., "Catastrophes: l'attribut sauvage", *Le Portique*, vol. 22, pp. 89–100, 2008.

[MOR 13] MOREAU Y., Catastrophes et mondes. Trajectoires et disputes du sens des aléas majeurs, Doctoral Thesis, Ecole des Hautes Etudes en Sciences Sociales, Paris, 10 October 2013.

[NOG 04] NOGUCHI T., *Ansei Edo jishin: saigai to seiji kenryoku*, Chimkuma Shobô, Tokyo, 2004.

[SMI 06] SMITS G., "Shaking up Japan: Edo Society and the 1855 catfish picture prints", *Journal of Social History*, pp. 1045–1072, 2006.

[SMI 07] SMITS G., "The 1855 Ansei Edo earthquake as a political event", *Columbia University Modern Japan Seminar*, New York, pp. 1–14, 2007.

[SMI 08] SMITS G., "Bakufu responses to the 1855 Ansei Edo earthquake", *Annual Meeting of the Association for Asian Studies*, Florida International University, Atlanta, 2008.

[STE 11] STEFFEN W., GRINEVALD J., CRUTZEN P., *et al.*, "The Anthropocene: conceptual and historical perspectives", *Philosophical Transactions of Royal Society (1938)*, pp. 842–867, 13 March 2011.

[UEX 56] UEXKÜLL J.V., KRISZAT G., *Mondes animaux et monde humain*, MÜLLER P. (trans.), Gonthier, Hambourg, 1956.

[WAT 11] WATSUJI T., *Fûdo, le milieu humain*, CNRS, Paris, 2011.

[WIT 65] WITTGENSTEIN L., *Le Cahier Brun*, Gallimard, Paris, 1965.

[WIT 92] WITTGENSTEIN L., *Leçons et conversations*, Folio, Paris, 1992.

List of Authors

Antoine ANDREMONT
IAME/INSERM
Université Paris Diderot
France

Thierry BERTHE
SCALE-CNRS/UNI
Rouen
France

Hélène BUDZINSKI
EPOC
Université Bordeaux 1
Talence
France

Vincent CATTOIR
CHU/UNI
Caen
France

Erick DENAMUR
IAME/INSERM
Université Paris Diderot
France

Filipe DUARTE SANTOS
Université de Lisbonne
Portugal

Murielle LAFAYE
CNES
Toulouse
France

Anniet LAVERMAN
METIS-CNRS
UPMC
Paris
France

Roland LECLERCQ
CHU/UNI
Caen
France

Véronique LOIZEAU
Ifremer/LBCO
Brest
France

Alexandre MAGNAN
IDDRI-Sciences.Po
Paris
France

André MARIOTTI
University Pierre and Marie Curie
Paris
France

Alain MIOSSEC
Institut de géographie
Nantes
France

André MONACO
Cefrem/CNRS-UPVD
Perpignan
France

Patrick MONFORT
CNRS/IRD/UNI
Montpellier
France

Serge MORAND
CNRS/IRD/UNI
Montpellier
France

Yoann MOREAU
Centre Edgar Morin
EHESS-CNRS
France

Kenny OBERLÉ
SCALE-CNRS/UNI
Rouen
France

Fabienne PETIT
SCALE-CNRS/UNI
Rouen
France

Jean-Charles POMEROL
University Pierre and Marie Curie
INSIS/CNRS
Paris
France

Patrick PROUZET
Ifremer
Issy-les-Moulineaux
France

Marie-Hélène TUSSEAU-VUILLEMIN
Ifremer/DS
Paris
France

Index

A, B

C

S, T

V, W, X

Other titles from

ISTE

in

Oceanography and Marine Biology

2014

Seas and Oceans Set coordinated by MONACO André, PROUZET Patrick
Ocean in the Earth System
The Land-Sea Interactions

MUTTIN Frédéric
Marine Coastal and Water Pollutions: Oil Spill Studies

2013

LÉCUYER Christophe
Water on Earth: Physicochemical and Biological Properties

Printed and bound by CPI Group (UK) Ltd, Croydon, CR0 4YY
13/09/2026
14961077-0005